A Future for Amazonia

A FUTURE FOR
AMAZONIA

Randy Borman and Cofán Environmental Politics

by MICHAEL CEPEK

University of Texas Press ⬥ Austin

Requests for permission to reproduce material from this work should be sent to:
    Permissions
    University of Texas Press
    P.O. Box 7819
    Austin, TX 78713-7819
    http://utpress.utexas.edu/index.php/rp-form

∞ The paper used in this book meets the minimum requirements of ANSI/NISO
Z39.48-1992 (R1997) (Permanence of Paper).

LIBRARY OF CONGRESS CATALOGING-IN-PUBLICATION DATA

Cepek, Michael.
    A future for Amazonia : Randy Borman and Cofán environmental
politics / by Michael Cepek.
        p.   cm. ⸻
    Includes bibliographical references and index.
    ISBN 978-0-292-73949-9 (hardback) — ISBN 978-0-292-73950-5 (paper)
    1. Cofán Indians—Politics and government.   2. Amazon River Region—
Environmental conditions.   3. Borman, Randall.   I. Title.
    F3722.1.C67C437   2012
    986.600498—dc23

                                                                    2012020598

doi:10.7560/739499

For Norm and Sibby, who gave me the first glimpse of the life I wanted to lead, and who did more than anyone else to make it happen.

Contents

Preface

I became interested in the Cofán people of Amazonian Ecuador because of the radical shift in their historical position. Their bodies and lands have suffered centuries of physical violence, environmental destruction, and political marginalization. For decades, observers predicted their "extinction." Against the odds, however, the Cofán became one of the most politically successful indigenous nations of greater Amazonia. Although I believe that their story is important for many reasons, I wrote this book to offer scholars, activists, students, and the larger public a motivating sense of optimism as they confront the tenuous situation of the world's cultural and biological diversity.

When I began my dissertation fieldwork in 2001, I felt alienated from much of academic anthropology. My colleagues were often critical, not only of each other but of many projects to make the world a more just and livable place. Whereas I appreciated their cautious stance toward attempts at intentional structural change, their perspectives appeared overly pessimistic. In many conversations, my hopefulness made me feel naïve. Slowly, I began to think that anthropologists were missing something. It was difficult to grasp the surprising turns of Cofán history from dominant stances. It was even more difficult to imagine writing about Cofán experiences in an optimistic tone.

I found partial reprieve in a submerged current of anthropological thought. Within a few years of the new millennium, three important volumes appeared at the borders of the discipline: David Harvey's *Spaces of Hope* (2000), Jonathan Lear's *Radical Hope: Ethics in the Face of Cultural Devastation* (2006), and Anna Tsing's *Friction: An Ethnography of Global Connection* (2004). Each book identifies "hope" as an essential element in the imagination, investigation, and realization of alternative futures. Although a trained geographer, Harvey is now a Distinguished Professor of Anthropology at the City University of New York, and liberatory social change is one of his long-held interests. Jonathan Lear, a philosopher and psychoanalyst at the University of Chicago, wrote *Radical Hope* about the human capacity for cultural renewal in the face of extreme devastation. And

Anna Tsing, a professor of anthropology at the University of California at Santa Cruz, continues her efforts to transform the discipline with *Friction*, which ends with a call for anthropology to embrace "utopian critique."

In *Spaces of Hope*, Harvey adopts the role of "insurgent architect" to argue that we should respond to troubled times not by falling back on Antonio Gramsci's dictum—"pessimism of the intellect, optimism of the will"—but by fostering an optimism of the intellect itself. He urges us to rely on more than blind hope as we struggle against the destructive dynamics of contemporary capitalism. Scholars, he asserts, must break free from the "political passivity, intellectual torpor and skepticism towards the future" (Harvey 2000:17) that undergirds the reigning pessimistic stance on the creation of new paths for social and political change. Like Ernst Bloch (1986), Harvey concludes that without some form of utopian imagination, we will be unable to produce alternative political realities.

In *Radical Hope*, Jonathan Lear explores a possibility that confronts all human beings: the ability to lose one's way of life, and thus all that provides a meaning and a telos to existence. His interlocutor is Plenty Coups, the last great chief of North America's Crow people. In a famous quote, Plenty Coups claimed that after the buffalo disappeared and Crow people moved onto reservations, "nothing happened." According to Lear's interpretation, Plenty Coups meant that the set of valued roles that gave Crow life significance no longer had validity. Rather than succumb to loss, however, Plenty Coups became a "new Crow poet" with the courage to live through a collapse of concepts. On the other side of devastation, Plenty Coups used a successful visionary experience to construct a vibrant social and political space in which Crow people could flourish in a new yet meaningful way. For Lear, Plenty Coups is an example from which all people can draw inspiration as they strive to overcome the specter of cultural loss.

In *Friction* (2004), Anna Tsing takes great pains to move beyond dominant trends in social and cultural theory. In her study of capital, knowledge, and activism on the island of Borneo, she emphasizes epistemic murk and political frustration. Nevertheless, she argues that her subjects have much to teach us about the creation of a more sustainable world. Although she does not deny the forms of governmental power that move within transnational environmental and human rights networks, Tsing suggests that scholars are prone to analytic exaggeration. She writes, "How has it happened that in order to stay true to hopes for a more livable earth, one must turn away from scholarly theory?" (2004:266). Rather than act as supporters, academics become skeptics of struggles to create local solutions and far-reaching coalitions. Tsing argues that when faced with overwhelming challenges, hope and a utopian sensibility are important elements of an approach that joins "generous" analysis to the practice of movements that can produce real transformations.

In different but compatible ways, Harvey, Lear, and Tsing urge us to make intellectual, ethical, and political commitments to the position that things could be otherwise. Even in situations where loss seems total and certain, there is the possibility that social, cultural, and political life will depart from standard narratives of predestined failure and inevitable destruction. In a similar vein, Roberto Unger suggests that adherence to the idea of a historical "script" is the key failing of social theory, which has never fully embraced the principle that people create the conditions of their lives in ways that are never totally determined, even in dire circumstances (1987). From the perspective of ethnography, David Graeber argues that even though anthropology maintains the broadest perspective on the openness of the human condition, its practitioners are too fearful of charges of romanticism and Orientalism to convince the public, as well as themselves, that another kind of world is possible (2004).

Other scholars from across the humanities and social sciences share a similar commitment to the notion of political possibility. With his characteristic dismissive wit, the philosopher of science and "symmetrical anthropologist" Bruno Latour (1993, 1999) argues that contemporary theorists make real enemies—including capitalist globalization, technological rationalization, and political domination—into total enemies. The false totalization gives impinging forces "fantastic properties," and it portrays people as powerless pawns of deterministic dynamics (1993:125). In his aptly titled book *In Defense of Lost Causes* (2009), the cultural critic Slavoj Žižek writes that the time of utopias is far from over. In a provocative move, he defends the Left's total politics of the past. Nevertheless, he does not argue for them in literal form. Rather, he justifies historical radical projects to demonstrate how uninspiring the existing liberal-democratic alternatives are.

Perhaps the scholar whose work most closely approximates the spirit of this book is the anthropologist Terence Turner. Turner has written for decades on the politics of South America's Kayapó people as well as the structure of contemporary capitalism. In a series of articles (Turner 1999a, 1999b, 2002, 2003b), he describes globalization as a successful project of elite actors. Nevertheless, he outlines a common ground for all who have lost power with political-economic shifts. Turner offers the possibility that subjugated groups, whether the once industrialized working class or such disenfranchised sectors as indigenous peoples, will recognize that they are engaged in a joint struggle to gain control of their collective means of self-production. Although the global political climate does not yet correspond to Turner's vision, he does allow us to imagine a coalitional platform that offers a path beyond fragmented conflicts, generalized acquiescence, and large-scale impotence.

Although they disagree on important issues, these thinkers are committed to an open and optimistic stance on the emergence of alternative futures. Whether pondering environmental destruction, cultural devastation, or political-economic oppression, they suggest that losing hope and the utopian spirit means losing the battle itself. Accordingly, they stress the need to construct accounts and theories that inspire and provoke rather than surrender and capitulate, either intellectually or politically. Following their example, this book tells an unlikely story of unexpected connections, surprising accomplishments, and far-reaching implications. Its key terms point to the mechanisms through which change is made: agency, creativity, imagination, and experimentation. If the book moves readers to consider the possibility that liberatory political and environmental realities may yet emerge, then it will have fulfilled its mission.

Acknowledgments

Many individuals and institutions helped this book to see the light of day. I researched and wrote it with support from the National Science Foundation, the Fulbright Commission, the Tinker Foundation, the Woodrow Wilson Foundation, the Mellon Foundation, the Field Museum of Natural History, the University of Chicago, Macalester College, and the University of Texas at San Antonio. I thank all of them for funding a long and often enjoyable journey.

Norm Whitten took me under his wing as an undergraduate at the University of Illinois, and he continues to be my most important interlocutor. His ethnographic imagination, ethical and editorial standards, and knowledge of all things Ecuadorian made it possible for me to work toward an understanding of Cofán life in Zábalo and Quito. As graduate advisors at the University of Chicago, Terry Turner, Manuela Carneiro da Cunha, and John Kelly were sources of support for nearly a decade. Terry taught me that a sound theoretical foundation is essential for ethnographic rigor and engaged practice. He gave me a great sense of what is possible as a scholar, a teacher, and an activist. Manuela inspired me with the breadth of her intellectual interests, her commitment to indigenous peoples' struggles, and her care and collegiality. She made the University of Chicago an amazing learning environment by bringing Amazonianists from Latin America and Europe to Haskell Hall. John initiated me into graduate study at Chicago. His eye for complexity and contingency gave me an appreciation for a kind of anthropology that always attends to the work of history and large-scale power formations.

Other scholars at the University of Chicago made my nine years as a doctoral student one of the most rewarding periods of my life. John Lucy, Michael Silverstein, John Comaroff, Sue Gal, and Nadia Abu El-Haj taught me about theory, ethnography, and the conceptualization of anthropological research. Jessica Cattelino, Judith Farquar, and Ray Fogelson were gracious participants in my dissertation defense. Eduardo Viveiros de Castro, Carlos Fausto, Philippe Descola, and Anne-Christine Taylor taught courses at Chicago. They allowed me to engage them in long conversations about the many ways to be an ethnographer of indigenous Amazonia. Anne Chien is the hub around which the Department of Anthropology revolves. Her miraculous ability to push through bureaucratic obstacles made life as a graduate student much more enjoyable (and productive) than it otherwise would have been.

Many of my most important colleagues studied with me at Chicago: Chris Ball, Anya Bernstein, Diana Bocarejo, Felipe Calvao, Kerry Chance, Biella Coleman, Bernard Dubbeld, Cassie Fennel, Santiago Giraldo, Jeremy Jones, Mark Koops-Elson, Genevieve Lakier, Paul Liffman, Tal Liron, Kathleen Lowrey, Sean Mitchell, Andrea Muehlebach, Clare Sammels, Stephen Scott, and Hadas Weiss, among others. I would like to give special thanks to Drew Gilbert and Paul Kockelman, two close friends whose intellectual companionship and emotional support continue to be lifesavers. Shane Greene, too, is a good friend and a helpful colleague who worked hard to coordinate a number of panels on which I was lucky to find a place.

Many other anthropologists provided critical aid during the writing of this book. I was fortunate to learn from the input of Glenn Bowman, Beth Conklin, Don Donham, Clark Erickson, Laura Graham, Jonathan Hill, Soren Hvalkof, Jean Jackson, Evan Killick, Stuart Kirsch, Eduardo Kohn, Carlos Londoño Sulkin, Flora Lu, Glenn Penny, Jean Rahier, Richard Reed, Stephen Rubenstein, Javier Ruedas, Fernando Santos-Granero, Amanda Stronza, Hanne Veber, Bill Vickers, and Paige West. My time in Ecuador with Michelle Wibbelsman, Dorothea Scott Whitten, and Diego Quiroga was essential. Chris Krupa continues to be a close friend and intellectual confidant, and it was a tremendously beneficial coincidence that our two years of dissertation fieldwork in Ecuador overlapped exactly.

I received research support from the Division of Environment, Culture, and Conservation at the Field Museum of Natural History. Debby Moskovitz, Dan Brinkmeier, Tyana Wachter, Alaka Wali, and Wendy Townsend allowed me to work with them in the field and to write about their involvements in Cofán conservation. They also funded my return trips to Cofán territory from 2003 to 2010. They are careful scientists and committed actors in the struggle for biodiversity conservation. Collaborating with them gave me an immense amount of hope that Amazonia's landscapes will receive the protection they deserve.

Many other people aided me in both Ecuador and the United States. Claire Nicklin and Andrew Reitz made my trips to Quito socially engaging and intellectually productive. Bub and Bobbie Borman were always ready to offer insights into Cofán culture and language. Ric and Ron Borman gave me another perspective on the history of their family. Freddy Espinoza and Maria Luisa Lopez made much of my research in Quito possible. Dayuma

Albán shared many insights from her work in Zábalo. Wade Davis pushed me to explore my interest in Cofán shamanism. John Himmelfarb allowed me to see some of the Borman family's roots in the greater Chicago area. Dan Trudeau and Paru Shah helped to hone my analyses during a post-doctoral fellowship at Macalister College. Jenn Andrades created a map of Cofán territory. And Elizabeth Joynes, Sadie Siviter, and Kimrey Hardin-Batts used their positions at the Foundation for the Survival of the Cofán People to help me in too many ways to mention.

My colleagues at the University of Texas at San Antonio have given me a nurturing intellectual home. I could not have asked for better coworkers than Sonia Alconini, Thad Bartlett, Kat Brown, Carolyn Ehardt, Jill Fleuriet, Dan Gelo, Jamon Halvaksz, Bob Hard, Jerry Jacka, Joanna Lambert, Laura Levi, and Jason Yaeger. In addition, I have learned a great deal from my students, especially the participants in my graduate seminars on cultural theory; economic anthropology; and culture, environment, and conservation.

It has been a pleasure to work with the University of Texas Press. I am extremely happy that Casey Kittrell and Theresa May took an interest in the manuscript. Casey has been a patient and insightful editor throughout the process of writing and revising. In addition, Michael Brown and Bret Gustafson provided superb commentary on the manuscript, helping me to make it into a much better book.

An important circle of friends gave me a strong dose of good cheer exactly when I needed it: Jerid Morris, Sean Peterson, Gerry Robledo, and Myriam Lopez. My parents, John and Charleen Cepek, supported me in many different ways during my graduate education and beyond. They instilled in me a sense of the value of intellectual commitment and political activism. More than anyone else, Amy Rushing has been with me through the highs and lows of the writing process. She somehow succeeds in convincing me that I'm a superstar, even when all evidence points to the contrary.

Most of all, I would like to thank the many Cofán people who worked as my tutors: Fidel Aguinda, Galo Aguinda, Toribio Aguinda, Aurelio Bustamante, Enma Chica, Angel Criollo, Atanasio Criollo, Claus Criollo, Emergildo Criollo, Floresto Criollo, Lorenzo Criollo, Rufino Criollo, Bolivar Lucitante, Basilio Mashacori, Luis Mendua, Valerio Mendua, Locrecia Queta, Aniseto Quenamá, Alfonso Yiyoguaje, Carlos Yiyoguaje, and Eduardo Yiyoguaje, among others. Various household heads allowed me to live with them in Doreno and Zábalo, and they also became important teachers: Antonio Aguinda and Saura Mashacori; Isolina Aguinda and Mimi Yiyoguaje; and Toribio Aguinda and Rebeca Mashacori. My ritual relatives in Zábalo, Doreno, and Dovuno continue to host me on my visits: Juanito Aguinda and Carmeza Lucitante (and Daicco); Alejandro Criollo and Rosia

Quenamá (and Roveto); and Jose Portilla and Vertina Lucitante (and Pilar). Many *Quito'su a'i* made my time at the Quito Cofán Center an enriching experience: Jonas Aguinda, Ramiro Aguinda, Felipe Borman, Federico Borman, Joshua Borman, Duglas Yiyoguaje, and Nivaldo Yiyoguaje.

A number of Cofán people deserve special mention for the crucial role they played in my research. Hugo Lucitante has become a great friend and an essential ally in both Zábalo and the United States. Mauricio Mendua worked tirelessly to help me understand Cofán social structure, cosmology, and shamanism. Roberto Aguinda accompanied me throughout Cofán territory as well as in Quito, and he was an excellent guide to the turbulent world of Cofán politics. Amelia Quenamá helped me to understand the many perspectives of a Cofán woman, and she taught me something new every day that I lived with her in Quito. Martin Criollo became a true collaborator in my work on Cofán history. And, of course, Randy Borman helped me in a multitude of ways. He placed a tremendous amount of trust in me when I was nothing more than a pesky undergraduate. I hope that he still thinks that I might be of some use in the Cofán experiment.

Finally, I wish to thank the journals that allowed me to reprint various portions of articles for this book. Parts of Chapter 3 appeared in "Bold Jaguars and Unsuspecting Monkeys: The Value of Fearlessness in Cofán Politics" (*Journal of the Royal Anthropological Institute*); parts of Chapter 4 appeared in "Essential Commitments: Identity and the Politics of Cofán Conservation" (*Journal of Latin American and Caribbean Anthropology*); and parts of Chapter 5 appeared in "Foucault in the Forest: Questioning Environmentality in Amazonia" (*American Ethnologist*).

Introduction

This book is a realist ethnography written in a utopian spirit. Its main characters belong to the Cofán ethnolinguistic group, an indigenous people of Amazonian Ecuador who face a microcosm of the forces that are devastating the world's cultural and biological diversity. After generations of dispossession at the hands of mestizo colonists, transnational oil companies, and Colombian armed factions, the Cofán nation, according to many observers' predictions, was in danger of disappearing and its rainforest territory of being destroyed. In the early 1990s, however, the Cofán began a process of mobilization that put them at the forefront of an expanding coalition between indigenous peoples and Western environmentalists. They are now the legally empowered, scientifically trained, and increasingly ambitious caretakers of more than one million acres of forestland.

On the basis of three years of ethnographic research in the lowland community of Zábalo and Ecuador's Andean capital Quito, this book explores Cofán people's innovative struggle for indigenous rights and environmental conservation. Although a number of developments explain the surprising reversal of Cofán prospects, a key factor in their success is Randy Borman, a Cofán man of Euro-American descent who has emerged as one of the world's most effective indigenous leaders. Blue-eyed and white-skinned, Borman was raised in a Cofán community by a pair of North American missionary-linguists. Armed with a Western education and fluency in English, Spanish, and A'ingae (the Cofán language), Borman has become a global media phenomenon. Magazine articles and television documentaries describe him as a "gringo chief" who combines North American whiteness, Amazonian indigenity, and committed political activism. Employing his intriguing persona and his intercultural capacities, Borman facilitates the Cofán nation's political struggles by constructing a creative set of partnerships with conservationist organizations, Western scientists, and the Ecuadorian state.

The introduction's subtitle refers to one of my central aims: to develop an "ethnography of possibility." With the phrase, I hope to convey multiple elements of my object and approach. The actors and efforts that compose Cofán politics point to new possibilities of cultural survival, environmental justice, and transnational collaboration in the twenty-first century. Cofán achievements are real. Nevertheless, they are undeniably provisional. Alliances, projects, and institutions change from year to year and moment to moment. They can be great successes one day, and they can disappear the next. The Cofán accomplishments I describe are a snapshot of possibility at a particular cultural, political, and historical juncture. The Cofán situation was different before my main fieldwork years of 2001 and 2002, and it changed after I left. Although I have attempted to keep track of the shifts with annual trips to Ecuador, a "final" account of Cofán politics is, by definition, impossible.

In addition to the conditional nature of Cofán achievements, a second sense of possibility motivates the book: the "experimental" nature of Cofán politics. When I first became interested in working with the Cofán, Randy Borman suggested that they would be happy to have me aboard "the Cofán experiment." The phrase refers to the shifting set of institutional forms into which Cofán people have thrown themselves: community conservation structures, ecotourism operations, partnerships with Western academics, formalized schooling projects, and the management of urban nongovernmental organizations (NGOs), among others. Cofán people's efforts profit from Borman's political creativity. They also reflect the flexibility of Cofán social structure, which allows individuals to embrace novel institutions without becoming anxious about their initial necessity or their ultimate consequences. Cofán people are more than willing to experiment with new ways of relating to each other, their environment, and encompassing political-economic forces. If results are beneficial and conditions are sufficient, the efforts continue. If not, they become material for revised goals and expectations.

Most importantly, I intend the book to communicate a third sense of possibility: the conviction that narratives of unstoppable cultural and biological loss are unfounded. I was drawn to Cofán politics because of Cofán people's triumph over tremendous challenges and terrible odds. Having worked, taught, and written on indigenous-environmentalist collaboration for years, I perceived a need to provide a story that is open-ended and optimistic rather than defined by conflict, failure, and an inability to see intercultural cooperation as anything more than a colonial project or governmental power. I believe that students, practitioners, academics, and activists—indigenous or otherwise—need hope if they are to devise creative solutions to the problems that face us. Although filled with complexity, the book provides reasons to believe in the possibility of indigenous empowerment and progressive conservation in the unlikeliest of times and places.

In short, possibility is the object, theme, and tone of the book. I tell multiple stories: social changes that emerge, endure, and transform; intentional experiments with the structure and context of a way of life; and an optimistic narrative on the fate of the world's cultural and biological diversity. To be sure, the Cofán struggle is one among many. From central Brazil to the Pacific Islands to postindustrial America, an array of efforts inspire a utopian sensibility: indigenous projects that combine new technologies with enduring lifeways; urban communities that play with novel forms of social and ecological organization; and a transnational, antiglobalization movement that is creating autonomous spaces in the fissures of contemporary capitalism. Viewed against this wider background, Cofán experiences hold valuable lessons for all actors who are striving to create a world whose difference is not doomed to destruction.

COFÁN LOSSES, COFÁN POSSIBILITIES

The community of Zábalo reflects a common history of Cofán loss and a shared field of Cofán possibilities. Attempting to escape the effects of oil pollution and uncontrolled colonization, Randy Borman and a small group of other Cofán people traveled far down the Aguarico River and founded the village in the late 1970s. It is located in the northeastern corner of Ecuador, along the heavily forested borders with Colombia and Peru. Zábalo residents hunt, fish, gather, and garden throughout their territory. Although some villagers speak Spanish, Quichua, Siona, and English with varying degrees of fluency, the majority consider themselves to be functionally monolingual in A'ingae.

Despite its remoteness, Zábalo is one of the most well-publicized indigenous communities in South America. During my time there, a film crew from the National Geographic Channel visited. They were preceded by others from *CBS America Tonight, Australian 60 Minutes,* and *A&E Investigative Reports.* In addition, Zábalo is the object of articles in *Life Magazine, Earth Island Journal, Condé Nast Traveler,* and *Cultural Survival Quarterly.* Joe Kane featured the community in *Savages* (1995), and Mike Tidwell made it the object of an entire book, *Amazon Stranger* (1996). In mid-2010, a journalist interviewed Borman (and myself) about political efforts that began in Zábalo. His research led to an article in the *Washington Post* and a story on National Public Radio.

The journalists who visit Zábalo are amazed by its strange conjunctures: English spoken alongside A'ingae, blowguns propped next to satellite telephones, and traditionally clad elders helping to enter ecological data into laptops powered by solar panels. What intrigues outsiders most, however, is Randy Borman. The Western media is fascinated by Borman's illustration of cultural distance, precisely because he exhibits key signs of "closeness" in a setting of radical alterity. The idea of "the gringo chief" strikes Western audiences as an index of the farthest flight or the deepest return, as well as the hybrid product of the passage itself. Many accounts write the story of Borman and Zábalo with archetypical images from colonial mythology. In Mike Tidwell's telling, Borman first appears in the observer's imagination as a stranger in the heart of darkness:

> [W]e heard the rumor about the small village down the river where the great white chief lived. The chief was an American, reportedly born and raised among forest Indians, a blowgun hunter since age four, a man gone totally native. With paint on his face and wild-boar eye teeth strung around his neck, this bushed-out Caucasian was leading the Indian campaign to keep the oil intruders out. The name *Kurtz* settled over my mind like equatorial heat when I heard this. I saw a malarial dream of a man. Conrad's antihero fast-forwarded to the late twentieth century. (1996:3–4)

Other accounts downplay evocations of remoteness to focus on Zábalo's symptoms of globalization. The journalist Joe Hooper writes, "There's no question but that Randy Borman has created an unusual place, a postmodern Indian village. Semioticians would have a ball with Sabelo [*sic*]" (1991). The political scientist and analyst of indigenous movements Alison Brysk suggests that, along with Rigoberta Menchú and Subcomandante Marcos, Borman is one of the most important faces of the clash between the "global village" and the "tribal village." In her words, "Randy Borman personifies the collisions brought about by globalization" (2000:4). In these perspectives, Borman's position looks less like that of Kurtz or Crusoe and more like the collapsed distance and difference of postmodernity's hybrid-man. Accordingly, the figures of Borman and Zábalo appeal to audiences as strange but intriguing icons of globalization's most progressive and promising currents, set against a background of extreme cultural and biological destruction.

Cofán History

Although they are important actors in the development of a Cofán-wide political project, the people of Zábalo form just one Cofán community among many. Cofán people live on both sides of the Ecuadorian-Colombian border, where the Andean foothills meet the Amazon rainforest. A'ingae,

Randy Borman in Zábalo (Photo by Michael Cepek)

which remains unclassified (Fischer 2007), is their primary language.[1] Current census figures suggest that approximately 1,200 Cofán reside in Ecuador's province of Sucumbíos with a similar number in Colombia's department of Putumayo (Fundación Zio-A'i 2000). In part, the national divide reflects a linguistic difference, with one dialect of A'ingae spoken along the Aguarico River and another along the San Miguel and other Colombian rivers. (The dialects are, however, mutually intelligible.) Due to intense political turmoil and social breakdown on the Colombian side, many Ecuadorian Cofán people do not consider their Colombian kin to be ethnic compatriots. As reasons for exclusion, they cite many Colombian Cofán individuals' inability to speak A'ingae, their abandonment of a forest-based lifestyle, and their long association with nonindigenous "killers," which has made them violent and unfit for peaceful community life.

Colonial and republican records do not consistently classify A'ingae speakers in the same group. People who today call themselves "Cofán" (sometimes spelled "Cofan," "Kofán," "Kofan," "Cofanes," "Kofanes," "Cufán," or "Copane" [Ortiz 1954]) were identified by other ethnonyms over the past

Cofán territory

four and a half centuries, including "San Miguel," "Sucumbíos," and "Macas" (Robinson 1979). In community contexts, most Cofán people refer to themselves as *a'i* (human being), which functions as a loose synonym for "Cofán." The early Jesuit missionary Rafael Ferrer estimated the total Cofán population in 1600 as 15,000. The Colombian geographer Juan Friede expanded the pre-Conquest estimate to 70,000 (1952). Smallpox, measles, polio, whooping cough, cholera, tuberculosis, and malaria drastically reduced the Cofán nation. The population hit a low point of fewer than 400 individuals after a 1923 measles epidemic. Thanks in part to the vaccination campaigns of Summer Institute of Linguistics/Wycliffe Bible Translator (SIL) missionary-linguists, which included Randy Borman's parents, the Cofán survived.

Until the mid-twentieth century, Cofán history is not well documented, although Juan Friede (1952),[2] Sergio Ortiz (1954), Scott Robinson (1979), Miguel Angel Cabodevilla (1996), Eduardo Kohn (2002), and Randy Borman (2009) do a good job of synthesizing the available materials. Very little is known about the pre-Conquest Cofán. A number of colonial sources make reference to unsuccessful Inca attempts to subjugate them in the late fifteenth and early sixteenth centuries (Borman 2009:223; Robinson 1979:24). Contemporary Cofán mythology makes no mention of Inca wars or relations with Andean peoples. Stories place much more emphasis on hostile relations with the Tukanoan and other peoples who continue to live to the east and south of contemporary Cofán territory, in Amazonia proper. Nevertheless, it is clear that A'ingae-speaking people occupied areas far into the Andean valleys in pre-Conquest times. Historical records suggest that

A'ingae may have been a common language in much of northern Ecuador's highland and montane regions before the Inca conquered the area's peoples prior to the arrival of the Spanish (Borman 2009:223).

In search of the "land of cinnamon," the military expedition of Captain Gonzalo Díaz de Pineda brought the Cofán into the Gobernación de Quijos in 1536. Upon finding evidence of alluvial gold deposits in the Aguarico and San Miguel regions, Spanish forces set up mining operations and towns (Robinson 1979:25). Hostile Cofán attacked and destroyed Spanish settlements in the 1500s. By the early 1600s, most of the mining centers were abandoned (Friede 1952). The first missionary activity occurred with the entrance of Pedro Ordóñez de Cevallos in the late 1500s. When Rafael Ferrer entered Cofán territory in 1602, he found that many of the Cofán "lords"—who appeared to lead relatively autonomous and very large settlements—had already acquired Catholic paraphernalia, such as statues, paintings, and bells (Borman 2009:224–225; Kohn 2002:550, 555). Contemporary Cofán tell myths about the miraculous powers of Catholic icons, which they describe as "God's Skin" and "Little God." Ferrer baptized 4,800 Cofán in seven years and founded a number of mission sites, some with as many as 3,000 inhabitants (Ortiz 1954). Unfortunately, Spanish soldiers and gold miners followed him into Cofán territory. The Cofán responded by drowning Ferrer in the Cofanes River in 1611 (Cabodevilla 1996:94–98; Friede 1952).

After the death of Ferrer, the historical record is extremely spotty. There are various references to killings of Jesuit priests and isolated "uprisings" in the seventeenth and eighteenth centuries, possibly in response to additional intrusions for gold (Friede 1952). The historical record contains very few references to Cofán people between the mid-eighteenth and the late nineteenth centuries. Friede uncovered evidence of large Cofán villages along the San Miguel River in the eighteenth and nineteenth centuries. Government forces destroyed the largest settlement in retaliation for the killing of a priest, and epidemic diseases apparently devastated the others (Friede 1952).

Even after the decline, Friede concludes that the Cofán who lived along the San Miguel and Guamués Rivers at the beginning of the twentieth century constituted a "large tribe." In the late nineteenth century, Ecuadorian officers estimated the number of A'ingae speakers along the Aguarico and Cuyabeno Rivers to be approximately 1,800 (Robinson 1979:29). Although the late-nineteenth- and early-twentieth-century Amazon Rubber Boom did not hit hard in Cofán territory, Capuchin priests entered in 1896 and caused significant disruptions. Starting in 1914, they rounded up Cofán people from dispersed communities and formed the new site of Nuevo San Miguel. In the settlement, the Cofán planted cash crops, attended school, listened to masses, and were "taught" to give up their long houses. In June of 1923, a measles epidemic ravaged the community. Survivors fled both upriver to the headwaters of the Aguarico and San Miguel and downriver to the Putumayo. Shortly thereafter, large-scale missionary efforts ceased (Friede 1952).

When the Aguarico River officially became part of Ecuadorian territory with the Colombo-Ecuadorian Boundary Commission in 1929, the overall Cofán population was at an all-time low of approximately 400 individuals, thanks to the measles epidemic and subsequent social disruption (Robinson 1979:32). Waves of nonindigenous colonists invaded Colombian Cofán territory. Although isolated colonists entered Ecuador, Cofán territory along the Aguarico was less affected, at least for a few decades. One family set up a small-scale gold mining operation near the community of Sinangoe in 1932 (Robinson 1979:33). A preliminary census by the SIL estimated the total Cofán population at 517 in the mid-1940s.

In 1955, Randy Borman's parents—Marlytte ("Bub") and Roberta ("Bobbie")—settled in the Cofán community of Doreno, where they raised their infant son and learned A'ingae so that they could produce a Bible translation. The Bormans also provided medical care, helping to cut off an emerging tuberculosis epidemic. Due largely to their interventions, the Cofán population began to recover. Carmelite missionaries, who began to work among the Ecuadorian Cofán in the second half of the twentieth century, later joined the medical effort. The Carmelites concentrated their programs in communities where the Bormans were not working, especially along the San Miguel tributaries and in Lago Agrio, which became the capital of the newly created province of Sucumbíos in 1991.

The Contemporary Political Situation

In the first half of the twentieth century, the Ecuadorian Cofán began their relationship with the force that would determine much of their lives through the new millennium: oil. In 1921, the Leonard Exploration Company, a front for Standard Oil, obtained a 2.5-million-hectare concession to explore and exploit deposits in Amazonian Ecuador (Martz 1987:46–47). The first all-weather road was constructed from the highlands to the lowlands in the 1930s, when the Anglo-Saxon Petroleum Company Ltd. (later revealed to be Royal Dutch Shell) acquired a forty-five-year 10-million-hectare concession (Schodt 1987:102). Nevertheless, the true invasion of Amazonian territory did not begin until a U.S.-based Texaco-Gulf consortium discovered large deposits of high-quality crude underneath Cofán territory and began commercial extraction in 1972 (Schodt 1987:105). Texaco-Gulf constructed a highway between Quito and the new provincial capital of Lago Agrio, which sat on top of the small Cofán village of Amisacho, which was inhabited through the 1960s.

Colonists flooded the lowlands both spontaneously and under government programs. After ten years of commercial extraction, half of the region's population was born outside of the lowlands (Serrano 1993:14–15). By 1990, two-thirds of Ecuador's Amazonian population was nonindigenous (Uquillas 1993:68). With petroleum-based development, Ecuador achieved the dubious honor of exhibiting the highest rate of deforestation in South America (Little 1992:76). Many Cofán people found it increasingly difficult to find enough game to sustain themselves on their shrinking patches of forest.

As the Cofán learned the consequences of land loss, they also began to suffer from the pollution caused by the dumping of billions of gallons of petroleum wastes into their environment. A series of studies note correlations between residence in Ecuador's oil-producing areas and a set of medical problems, including increased rates of malnutrition, cancer, birth defects, developmental disorders, miscarriages, skin and respiratory ailments, and diarrhea (Centro para Derechos Económicos y Sociales 1994; Hurtig and San Sebastián 2002, 2004; Kimerling 1991; San Sebastián, Armstrong, and Stevens 2002; San Sebastián and Hurtig 2004). Texaco drilled one of its most productive wells inside the boundaries of Doreno. The well—"Dureno Uno"—produced more than one million barrels of oil. In return, the Doreno Cofán received the destruction of their forest, decades of toxic runoff, and three spoons, which early oil workers gave them as "gifts" (OINCE 1998).

After the initial decades of shock and incomprehension, the Cofán nation began to resist the consequences of oil extraction. In the late 1970s, Borman and a group of Doreno residents decided to flee the effects of petroleum-based development by forming Zábalo. In 1987, Zábalo residents joined their Doreno relatives to block a road that Texaco attempted to build through their community, which was already an island of forest in a sea of colonist settlements. In 1991, seismic crews began working in Zábalo's territory. After initial Cofán resistance, the state oil company, Petroecuador, succeeded in drilling two exploratory wells near the village, without Cofán permission or proper state permits. In 1993, the people of Zábalo detained workers, confiscated equipment, took over a well, and burned down a heliport to force Petroecuador out.[3] On October 12, 1998, a Cofán-wide contingent took over and shut down Dureno Uno, the most visible sign of oil exploitation in Cofán territory.

In addition to direct actions, the Cofán are plaintiffs in *Aguinda v. Chevron*, a precedent-setting, multibillion-dollar class-action lawsuit. Lawyers first filed the case in the United States on behalf of the thousands of indigenous and nonindigenous Ecuadorians who have suffered the environmental devastation caused by oil extraction (Berlinger 2009; Vickers 2003:60–61; www.chevrontoxico.org 2010). They made creative use of the U.S. Alien Tort Statute to base the case in the New York district where Texaco's headquarters are located. After rounds of appeals, a federal judge sent the case back to Ecuador. After Chevron acquired Texaco in 2001, North American environmental NGOs brought Cofán leaders to Chevron's California headquarters to communicate their experiences and make their demands. In 2011, an Ecuadorian judge ruled in the plaintiffs' favor to the tune of $18 billion. Nevertheless, both parties are appealing the decision, and it is unclear when a final ruling will occur.

As oil prices surge, and as companies from all over the world acquire concessions in northeastern Ecuador, more conflicts will emerge. Currently, the political-economic pressure on the Cofán is substantial. Oil is Ecuador's most important export. Its revenues cover nearly half of the federal budget. During the first decade of the twenty-first century, Venezuela was the only South American country that exported more oil to the United States than Ecuador, and Ecuador was the third-largest supplier of oil to the West Coast, after Saudi Arabia and Iraq (Bremmer 2006). By 2011, Ecuador had become an important supplier to China (EFC News Services 2011). In 2009, Petroecuador opened a new oil field directly upriver from Zábalo, although it has yet to cross the community's boundaries.

The petroleum industry has also destroyed Cofán territory on the Colombian side of the border. The Colombian Cofán are widely known as one of the indigenous peoples most affected by oil and colonization. Environmental destruction in their homeland is extreme: over 70 percent of the Guamués valley is deforested, and nearly half of the forest in the district of San Miguel is gone (Vargas 2004:117). Only a small minority of Colombian Cofán continue to practice a hunting, fishing, and horticulture-based lifestyle. At the beginning of the millennium, twenty-eight multinational companies were extracting oil from the department of Putumayo (Edeli and Hurwitz 2003:14). Although the environmental destruction is substantial, an equally pressing problem is the money that oil companies pay to armed groups for "protection" (Ramírez 2002). Colombia's militant factions, however, have additional sources of revenue.

Putumayo is the epicenter of Colombia's cocaine trade, which produced $3.5 billion nationally and $46 billion globally in 2002 (Adelman 2002:55). In the first decade of the twenty-first century, two of the groups who profited most from drug money were the leftist Revolutionary Armed Forces of Colombia (FARC) and the right-wing paramilitary United Self-Defense

Forces of Colombia (AUC). Although the FARC and other guerrillas such as the National Liberation Army (ELN) have deep roots in decades of leftist politics, the late 1990s ushered in a different phase of conflict. During the main years of my fieldwork, Colombia had the highest murder and kidnapping rate in the world (Hylton 2003:52). In 2002, there were over twenty murders a day. Homicide became the leading cause of death of men and the second leading cause of death of women (Youngers 2004:141).

As part of its war on drugs, the United States funded the intensification of the Colombian conflict. Contemporary U.S. policy took form in 1989 with George H. W. Bush's "Andean Initiative." Bush wanted to solve the United States' drug problem by attacking its "source"—i.e., coca and poppy production centers in Andean countries. Bill Clinton followed Bush's policy. In 2000, 65 percent of the federal antidrug budget was directed to the "supply side." Most of the funds went to military and police programs in drug-producing countries (Youngers 2004:130). In 2002, George W. Bush took his father's cue and created the "Andean Regional Initiative," which focused on financing strategic military interventions in Colombia, Ecuador, Bolivia, and Peru (Adelman 2002:70; Youngers 2004:130–131). The Bush administration cast the Colombian conflict as a battle against terrorism. After September 11, both the FARC and the AUC made the United States' terrorist list.

The Colombian civil war intensified in 2000 with President Andrés Pastrana's "Plan Colombia," or "Plan for Peace, Prosperity, and the Strengthening of the State." President Clinton's original contribution of $1.3 billion made Colombia the United States' third-largest recipient of military aid, after Israel and Egypt. Supported by dozens of U.S. Black Hawk and Super Huey helicopters as well as training and firepower, the Colombian military pushed into Putumayo in December of 2000 to fumigate coca fields with a toxic mix of Cosmo-Flux 411F and Ultra Glyphosate (Roundup Ultra). The FARC responded by increasing its military training and investing in handheld antiaircraft weapons. In March of 2002, the Colombian army invaded the "safe haven" that Pastrana had ceded to the FARC as part of peace negotiations. Violent encounters between the FARC, the AUC, the Colombian military, and civilians increased rapidly. Many of the involved actors extended their sphere of operations into Ecuadorian territory. In 2002, the Colombian people elected the ultraconservative Álvaro Uribe Vélez as their president. He declared that the only way to end the civil war was to beat the FARC through unrestrained military might. Uribe's policy significantly reduced the power of the FARC, although it is still a dominant actor in Cofán territory.

Colombian indigenous leaders began negotiating with the FARC and the ELN in the mid-1980s to keep the conflict out of their communities (Edeli and Hurwitz 2003:14–15). Nevertheless, the leaders admitted that over six thousand indigenous youths joined the guerrillas while a much smaller number joined the paramilitaries (Richardson 2003:19). On the ground, the situation is complex. Many Colombian Cofán began cultivating coca in the 1980s (Ramírez 2002:136). In 2001, the U.S. Agency for International Development (USAID) agreed to give the Colombian Cofán $864,000— $1,774 per family—to manually eradicate their coca crops (Fletcher 2003:22–23). In December of 2000 and January of 2001, paramilitaries assassinated three Colombian Cofán leaders. In response, approximately eighty Cofán families crossed into Ecuador to seek refuge with distant kin, but the Ecuadorian Cofán accepted very few of the newcomers (Vickers 2003:67–68). They feared that the Colombian Cofán had lost their "Cofán heart," thereby becoming thieves and murderers. Eventually, the great majority returned to Colombia. None stayed in Zábalo.

Ecuador's border province of Sucumbíos, where 90 percent of the population lives in poverty (Edeli and Richardson 2003:66), has been completely transformed by the Colombian conflict. As early as 1993, the Ecuadorian army clashed with FARC combatants (Vickers 2003:64). Over the last two decades, Ecuadorian soldiers have discovered hundreds of coca fields, coca processing plants, and guerrilla camps in the province. Colombian guerrillas and paramilitaries now have Ecuadorian offshoots. In the first six months of 2002, there were more than one hundred murders in Lago Agrio, which had 25,000 inhabitants at the time (Edeli and Richardson 2003:67).

The herbicides from Colombian coca eradication drift over the border. Ecuadorians report that the chemicals destroy their subsistence crops and cause skin rashes, fevers, headaches, diarrhea, and even death (Vickers 2003:66). At least three indigenous communities on the Ecuadorian side of the border have been abandoned because of paramilitary threats. The Ecuadorian military has tortured members of the same communities for information (Walcott 2003:58–59). High-profile kidnappings of oil company workers in 1999 and 2000 destroyed the ecotourism business in Sucumbíos, which was the central source of income for many Cofán people. Thousands of Colombian refugees entered Ecuador. In order to deal with the problems, the Ecuadorian government created a "Plan of Internal Defense" (Vickers 2003:67). It made use of $12 million of U.S. military aid to augment its border security with 12,000 troops (Edeli and Richardson 2003:67–68).

Although Ecuadorian Cofán communities have experienced little direct violence from either guerrillas or paramilitaries, it is possible to hear warplanes and explosions from some villages. During the highly violent years of

2001 and 2002, the residents of Zábalo avoided making trips to Doreno and Lago Agrio, which were close to the most dangerous zones. For a short time, the military installed a checkpoint in the colonist town across the river from Doreno, through which arms and drug production materials pass. Although the troubles usually seemed distant during my fieldwork, one morning in 2002 a burlap sack containing human body parts washed up in Zábalo on the shore of the Aguarico River. No one knew its origin, but everyone concluded that it must have had something to do with Colombia.

Zábalo residents have heard many stories about paramilitary atrocities, including the use of chain saws to dismember their live enemies and the consumption of their enemies' flesh. Many Cofán refer to paramilitaries as *cocoya*, the same word they use for malevolent supernatural agents. Although travels across the border have decreased because of the violence, a number of Zábalo residents continue to visit Colombian communities in order to spend time with kin and to seek the treatment of powerful Cofán shamans.

Caught between the consequences of oil exploitation and civil war, the Cofán have received inconsistent aid from the Ecuadorian state, which has suffered severe instability. Ecuador went through seven presidents in a decade. Seventy percent of its population lives in poverty, and nearly 10 percent of its citizens have gone abroad to find work (Whitten 2003:20). In 2003, emigrant remittances surpassed every source of national revenue except petroleum (Whitten 2004:452). The country appears to be increasingly polarized between sectors that desire to implement neoliberal reforms and U.S.-backed policy objectives and the populist, nationalist, and left-wing segments that want to steer the country in a new direction.

One of the most important social sectors advocating radical change is Ecuador's indigenous population, which comprises as much as 40 percent of the country's 14 million citizens. Approximately 95 percent of Ecuador's indigenous people are native to the Andean highlands, whereas much smaller groups live in the coastal and Amazonian provinces (Collins 2004:39; MacDonald 2002:176). For more than fifty years, Ecuador's indigenous peoples worked to build Latin America's most powerful indigenous movement. Beginning in the Amazonian region with the formation of the Comuna San Jacinto del Pindo in the mid-1940s (Whitten 1976, 1985) and the Federation of Shuar Centros in the early 1960s (Salazar 1981), Ecuador's indigenous peoples have created organizations for nearly all of the country's peoples. Ethnic federations are represented by regional organizations: the Confederation of Indigenous Nationalities of Amazonian Ecuador (CONFENIAE), the Confederation of Peoples of the Quichua Nationality of Ecuador (ECUARUNARI, based in the highlands), and the Confederation of Indigenous Nationalities of the Coast (CONAICE). Created in 1986, the Confederation of Indigenous Nationalities of Ecuador (CONAIE) represents all indigenous organizations at the national level.

Ecuador's indigenous peoples have consistently challenged state policies that threaten to leave them susceptible to the whims of national and transnational capital by weakening their land base and limiting their powers of self-determination. The indigenous movement has been at the forefront of tremendous collective efforts: the Levantamiento (uprising) of 1990; the Caminata (march) from the Amazonian lowlands to Quito in 1992; mobilizations against neoliberal reforms of the Ley Agraria (Agrarian Law) in 1994; participation in drafting the 1998 constitution (which declared Ecuador a "pluricultural and multiethnic state"); an important role in the coup of January 21, 2000; and leadership of collective efforts against state policies and corporate actions every year since the new millennium began.

The Ecuadorian Cofán began developing an ethnic federation in the early 1990s. Nevertheless, they did not succeed in legalizing their organization—the Federación Indígena de la Nacionalidad Cofán del Ecuador (Indigenous Federation of the Cofán Nationality of Ecuador; FEINCE)—until 2001. From the perspective of the people of Zábalo, a more important institution is the Foundation for the Survival of the Cofán People (FSC). Randy Borman and other Cofán leaders established the FSC in 1998 as an NGO that could operate apart from the national indigenous movement. Many Cofán activists have long felt distanced and manipulated by CONAIE and CONFENIAE, as the larger federations are dominated by much more populous ethnic groups.

With the election of President Rafael Correa in 2006, the Cofán appeared to receive a new ally. Correa, who holds a Ph.D. in economics from the University of Illinois, is part of Latin America's new left leadership. He has been Ecuador's most popular president in decades. Soon after taking office, he personally signed a contentious bill authorizing land acquisition for the Cofán nation in the Cofanes River region. In 2008, he organized a constituent assembly to write a new constitution, which passed a popular vote by a wide margin and included substantial rights for indigenous peoples and "Nature" itself. In the mining and oil sectors, Correa has worked hard to wrest control and profits from transnational corporations through a series of asset seizures and contract renegotiations. Much of the additional revenue is destined for new social programs (Swartz and Álvaro 2010).

By 2008, however, it became clear that Correa's increasing commitment to expanding mining and petroleum production put him in direct confrontation with the country's indigenous peoples, many of whom oppose a development model focused on extractivism. The government has imprisoned indigenous leaders on charges of "terrorism" for protesting Correa's policies. In addition, Correa has threatened expulsion for many activist *"gringuitos"* ("little gringos," or North Americans), whom he accuses of manipulating the country's indigenous and environmental movements (Webber 2010). In June of 2010, CONAIE felt so alienated from the Correa administration that it created its own parallel "Government of Peoples and Nationalities." Although Cofán people are not directly involved in these developments, Correa's commitment to increasing oil revenues at almost any cost represents a significant danger for their future.

Devastation and Possibility

The history of Cofán loss is extreme. Before the Spanish arrived, Cofán forebears were embroiled in battles with the Inca Empire. Shortly after the Spanish Conquest, waves of colonial forces descended into Cofán territory, bringing wars and epidemic disease. Centuries of missionary attempts continued the dynamics. Only a small fraction of the original Cofán population survived the first few decades of the twentieth century. When the oil industry made Cofán territory the epicenter of its operations in the 1960s, events took a turn for the worse. Colonists expropriated Cofán land, companies polluted forests and rivers, and the Ecuadorian state showed little interest in the survival of a tiny remnant of what was once a powerful native nation. The Cofán hung on to only small fragments of their land. By the beginning of the new millennium, the spillover of violence from Colombia's civil war and drug trade made their situation even more precarious. Decades-old predictions of Cofán ethnocide appeared to become more and more realistic with each additional crisis.

The Cofán nation, however, did not disappear. In the 1970s, two of the communities in the most colonized areas—Doreno and Dovuno—received legal title to a small portion of their territory. Shortly thereafter, the community of Sinangoe established a legalized land base. With the state's recognition of Zábalo in the early 1990s, the Cofán nation's lands expanded exponentially. In the new millennium, Cofán leaders worked through the Ministry of Environment to achieve control of even more territory. Today, Ecuador's Cofán population holds rights to 433,400 hectares of Andean and Amazonian forests. As a result of increasingly strong relations with scientific institutions, the international donor community, and some sectors of the Ecuadorian state, Cofán people have achieved a rate of zero deforestation in their territory. They are using their land base to attain new levels of status, power, and wealth at national and global levels.

This book examines the radical reversal of Cofán prospects at a crucial juncture in Cofán history. The work of Randy Borman and the residents of Zábalo has been an essential element of the unlikely success story. Cofán people have struggled to overcome their challenges by integrating novel sociocultural forms—nongovernmental organizations, science and conservation partnerships, large-scale enforcement mechanisms, and urban schooling programs—into a way of life that continues to be grounded in the forest environment. With a new set of allies, institutions, and capacities, Cofán people are restructuring their society to advance their central project: escaping their position as the unpaid and uncredited objects of scientific expertise, state power, and transnational conservation work to become their own fully empowered agents. In so doing, they are laying the foundation for the expansion of their way of life into the next century. Moreover, they are providing a case from which we all can derive inspiration as we ponder the possibility of a world in which cultural and biological diversity are not doomed to destruction.

A FEW WORDS ABOUT LITERATURE, THEORY, AND METHOD

The Cofán experiment is a unique political project. Nevertheless, it involves many of the same issues that other indigenous peoples are negotiating. Scholars have written extensively on the set of overlapping topics, which include: relations between indigenous peoples and Western environmentalists (Brosius 1997, 1999a, 1999b; Buege 1996; Carneiro da Cunha and Barbosa de Almeida 2000; Chapin 2004; Conklin and Graham 1995; Lohmann 1993; Lu Holt 2005; Nadasdy 2005, 2006; Redford 1991; Redford and Stearman 1993; Varese 1996; West 2006); the cultural and political dynamics of indigenous activism (Brysk 1996, 2000; Conklin 1997; Friedman 1996; Hale 2006; Jackson 1995; Jackson and Warren 2005; Li 2000; Muehlebach 2001, 2003; Ramírez 2002; Ramos 1994, 1998; Rogers 1996; Warren and Jackson 2002); the work of indigenous representatives and movement intellectuals (Brown 1993; Greene 2009; Gustafson 2009; Lauer 2006; Rappaport 2005; Veber 2007; Viatori 2007); the ethnoracial difference of Amazonia's "messianic" leaders (Bodley 1972; Brown 1991; Brown and Fernández 1995; Carneiro da Cunha 1973; Clastres 1978; Crocker 1967; Hugh-Jones 1996; Pereira de Queiroz 1969; Santos-Granero 1992; Santos-Granero and Barclay 1998; Shapiro 1987; Veber 2003; Wright 2002; Wright and Hill 1986, 1992); and the application of the concept of "governmentality" to processes of environmental management and other forms of directed change (Agrawal 2005a, 2005b; Braun 2000, 2003; Darier 1999; Escobar 1994; Ferguson 1994; Fisher 1997; Li 2007; Luke 1999;

Moore 2000; Rutherford 1999; Sawyer 2004). I grapple with these issues in a number of articles, which deal explicitly with the anthropological literature (Cepek 2008a, 2008b, 2009, 2011). For the sake of wider readability, however, I attempt to keep this book relatively free of specialist debates and theoretical conversations. I believe that the story of Randy Borman and Cofán political projects will be of interest to many people, and I do not want to focus too heavily on the specific concerns of an academic audience. Nevertheless, a few short remarks about my theoretical and methodological positions are necessary.

The main topic of *A Future for Amazonia* is the pragmatics of Cofán political mobilization. With the word *pragmatics*, I intend to position my work within traditions of philosophy and social theory that view sensuous activity as the most basic level of human being. I take inspiration from multiple theoretical schools: the activity-, production-, and value-based approach of Karl Marx and a number of contemporary anthropologists (Graeber 2001; Kockelman 2010; Munn 1986; Turner 1984, 2003a, 2008); the American pragmatism of George Herbert Mead (1934) and John Dewey (1929); the phenomenology of Martin Heidegger (1996) and Alfred Schutz (Schutz and Luckmann 1973); the sociopolitical thought of Roy Bhaskar (1989, 1998), Cornelius Castoriadis (1998), and Roberto Unger (1987); and the sociology of Peter Berger and Thomas Luckmann (1966), Pierre Bourdieu (1977), and Hans Joas (1996, 2000).

In colloquial usage, *pragmatic* conveys the notion of adaptability. In a similar sense, theoretical pragmatism's central concepts are uncertainty, problem solving, socially distributed knowledge,[4] and experimental intelligence and action. The last idea suggests a number of principles: people act purposively on the basis of imperfect information; they attend actively to the often unintended consequences of their actions; and they are fundamentally open to revising their modes of acting in and on the world. According to the general approach, a key aspect of creative human action is the interruption of unreflected habit in practical circumstances. Disrupted actions and troubled understandings provide the conditions for problem solving, knowledge expansion, and self and society transformation. In my usage, pragmatism also depends on a realist orientation. From this perspective, all action depends on a socionatural milieu that people confront in the taken-for-granted form of the "everyday life-world" (Schutz and Luckmann 1973:3).

I use the basic aspects of a pragmatist approach to conceptualize political mobilization as: (1) the processes by which a collectivity develops a value-based sense of itself in a situation of crisis; and (2) the processes by which the collectivity employs that sense to become the agent of its own transformation. Political mobilization is a form of intentional social change. As people work to transform themselves and their world, they are enabled and constrained by current sociopolitical circumstances and enduring cultural orientations. I conceive "culture" according to Bourdieu's perspective (1977): as a loosely organized, constantly transformed, and complexly overlapping set of schemes of perception, interpretation, appreciation, and action.

For the residents of Zábalo, to identify as Cofán is to participate in a calm, symmetrical, and fulfilling style of sociality in a community of familiars. Cofán people experience the disruption of preferred forms of sociality in relations with ethnic others, whom they associate with anger, violence, conflict, and selfishness. Accordingly, the Cofán view their current sociopolitical situation—of invading colonists, a disastrous oil industry, and Colombian warfare—as the ever-greater encroachment of disruptive difference, which threatens their ability to lead a meaningful and satisfying life.

Cofán people's difficulty in practicing a valued way of life constitutes their contemporary situation as a generalized crisis. Articulations with ethnic others, however, are not just matters of defense. Cofán people now consider resources that originate outside of their communities to be essential to a stable and satisfying existence. Their political action aims to transform their society and its external ties so that they can keep violent difference at bay while extracting the necessities that outside domains offer.

My account of the pragmatics of Cofán political mobilization is related to two analytic approaches to the contemporary situations of Amazonian peoples. On the one hand, Amazonia is still a region where the "classical" study of indigenous culture and social structure is an important intellectual movement. Led by Brazilian, French, and British anthropologists and partially inspired by Claude Lévi-Strauss's structuralism, many Amazonianists construct fine-grained, people- and place-based ethnographies that have a strong cosmological focus but are largely holistic in nature, covering all aspects of native society. For neoclassical Amazonianists, understanding indigenous culture is an end in itself. On the other hand, a different set of scholars, based mainly in North America, focus on the contemporary political situation and global activism of indigenous Amazonians. Their studies profit from a closer attention to political and poststructural theory as well as anthropology's "critical moment" in the 1980s and 1990s. Their methods typically involve a kind of multisited ethnography that does not place the same emphasis on building an encyclopedic portrait of indigenous culture, as developed by a long-term study of a single people or extensive work in a native language. In short, the former emphasizes culture and continuity, sometimes at the expense of history and politics. The latter, in contrast, stresses global connections and political-economic entanglements.

Nevertheless, it pays less attention to the nuances of cultural difference that structure the experience of all people, Amazonian or otherwise.

I attempt to straddle the divide between the two intellectual styles. Rather than explain my position with an extended theoretical discussion, however, I would like to illustrate my stance with an example of the kind of analysis that composes the book. In Chapter 5, I explore scientific conservation in Zábalo. While speaking to Zábalo residents, I tried to understand how working on conservation is different from other modes of experiencing the forest, as well as how it stimulates people to reflect on their global positioning. Many of my consultants contrasted project work with less structured subsistence activities, which they described approvingly with the word *opatssi*. The word is the central term in Cofán value discourse, and it refers to a calm and satisfying style of community-based conviviality. For many individuals, I discovered, scientific work is anxious and therefore decidedly non-*opatssi*, mainly because of its structured form and the conflicts it creates, which take the shape of disruptive jealousy and anger.

My consultants' discussions of the term *opatssi*, however, contrasted the socioexistential state to which it refers with the moral stance and practical duties of a shaman. Consequently, I directed my attention to an unplanned topic as I tried to figure out how participation in scientific conservation is similar to shamanic practice. My studies of shamanism led to questions about the history of Cofán warfare and leadership as well as dietary taboos, rites of passage, and menstrual practices. In my investigation of Cofán activism, I was already devoting much of my fieldwork to the site of Zábalo and the language of A'ingae. Nevertheless, learning how to brew and consume hallucinogens, as well as wrapping my mind around the infinite complexities of Cofán cosmology, was not what I imagined when I set out to study "indigenous politics in the twenty-first century." Nevertheless, the investigation was essential. Cofán culture exists, and I began to understand how it structures political action, interethnic collaboration, and everyday community life.

From my perspective, holistic ethnography's particular contribution stems from two methodological absolutes: long-term copresence as a mode of access to an object, and radical openness to topic formation during the study of that object. Regardless of scholarly objectives, extended immersion in settings whose distinctness we presuppose allows for an investigative perspective that reaches our deepest levels of bodily experience and theoretical expertise. When we push ourselves to learn new languages, to obey a different logic of life, and to do without practical familiarity, we activate new critical and creative faculties. Extended immersion in alien contexts destroys doxa. If we can deal with the inevitable breakdowns, waves of reflexivity allow us to think critically about the sociocultural resources composing the lives of the people with whom we work as well as our own.

The second methodological principle involves a central ethnographic access point: the community setting. Joking with colleagues, I often say that I could not escape my object during fieldwork. It was always all around me, constantly knocking on my door, feeding me, and waking me up. Many anthropologists now have to search out their objects in an office space, an Internet message board, a newspaper archive, or a delimited political event. Thanks to my constant presence in Zábalo (as well as the Quito Cofán Center, which is also a kind of community), I was able to shadow the great majority of my informants through familiar routines, troubled circumstances, and singular events. My research style allowed me to view a variety of phenomena against the continuous background of everyday, shared life.

The point is not that communities are closed or timeless. Rather, for such peoples as the Cofán, they overlap with what a phenomenologist would call a "lifeworld." Cofán people spend the great bulk of their time in their communities; these communities include the majority of their significant others, they are the primary sites of socialization, and their maintenance is the central aim of Cofán activists who depart from them in order to sustain them. By situating myself in a community setting, my main concern was not a presupposed topic, but a phenomenologically construed ground—a perspective. Because I was able to be maximally present in this ground, I could attempt to disclose objects from the position of a socioculturally distinct pragmatic center, even if my main concern of political mobilization looked fragmented, partial, or even nonexistent from the view I gained.

The theoretical and methodological differences of anthropologists who work with "traditional" and "nontraditional" topics are a matter of degree, not of kind. Degree, however, is important. The objectives, methods, and political possibilities of embedded, holistic ethnography are among the lasting contributions of the discipline. This is especially true now, as we are becoming ever more aware of how mindful immersion in contexts of difference allows us to develop novel perspectives on the powers that structure our lives.

I remain committed to ethnography's promise of uncovering the social, historical, and semiotic chains that constitute the meaning of any event, object, or action, whether shamanic rituals in Zábalo or intercultural relations with scientists, government officials, and urban schoolteachers. Without understanding the mediating work of culture in practice, we will never be able to construct convincing accounts of the lived dynamics of a globalizing world. My argument might seem commonsensical to most anthropologists. The approach I advocate, after all, is not a new one. I take the lead of scholars who analyze global political action through the lens of long-term, open-minded, people- and place-based ethnography. From my perspective, the works of the Amazonianists Laura Graham (2002, 2005), Terence Turner (1995, 2000), Hanne Veber (1998, 2007), Norman Whitten (1976, 1985), and Norman and Dorothea Scott Whitten (2008) are especially compelling in this regard. I do my best to follow their example.

FIELD SITES AND RESEARCH HISTORY

Randy Borman was the first Cofán person I met. I interviewed him in 1994 while conducting research in Quito for my bachelor's thesis on Cofán resistance to petroleum-based development (Cepek 1996). After enrolling in the doctoral program at the University of Chicago in 1997, I completed a proposal to investigate the history of the oil industry from the Cofán perspective at Doreno. I traveled to Doreno for one week in 1998 and one month in 1999. Toribio Aguinda, president of the Cofán ethnic federation, hosted and encouraged me. Two weeks before I was to depart for two years of Fulbright-supported research in Ecuador, however, I received a call from the Ecuadorian head of the Fulbright Commission. She told me that I could no longer do my study. The U.S. State Department had informed her that Doreno was too dangerous for a lone gringo, given the guerrilla, paramilitary, kidnapper, and coca-grower presence in the area. I spoke to Borman, who invited me to stay with him in Quito, from which I could make trips to the lowlands, as circumstances permitted.

Although I was able to travel to Doreno multiple times in 2001 and 2002, I became interested in the political projects that were occurring between Quito and Zábalo. Borman and other Zábalo residents graciously agreed to let me poke my nose around their homes and lives for an initial period of two years. While living in Zábalo and Quito, I learned a tremendous amount about the cultural perspectives and political conditions that unite the Cofán nation. Nevertheless, my extensive familiarity with the two sites gave me a particular vantage point. The contemporary Cofán population is extremely diverse. Differences occur along nation-state borders, dialect groupings, community affiliations, and individual personalities. Although I eventually had opportunities to work with people from all Cofán communities—and to temper my conclusions accordingly—readers should remain cognizant of the fact that I am most familiar with the situation in Ecuador, and especially in Zábalo.

Zábalo is located along the Aguarico River in northeastern Ecuador. In 2010, approximately 175 people lived there. Its territory borders Peru along the Lagartococha River, and it is less than a day's walk to Colombia from the community's northern limit on the Güeppí River. Within the central settlement, *naccu* (extended family unit) membership structures residential patterns. Visiting and sharing networks are especially strong within *naccu* and household clusters, although they also span the entire community. The community holds meetings and collective work parties in its main settlement, which is also the home of the community canoe, outboard motor, radio, school, soccer field, generator, and tourist huts. Zábalo residents hunt, fish, and gather throughout their territory. They maintain horticultural fields next to their houses and in more distant locations along the Aguarico.

Although Zábalo began its formation as a Cofán community in the late 1970s, it was not until 1992 that the Ecuadorian state recognized its existence. Two earlier attempts to secure title to nearly 40,000 hectares of land—the first with the aid of CONFENIAE in 1986 and the second with the aid of the still-unofficial Cofán federation in 1990—bore no fruit with the Ecuadorian Institute of Agrarian Reform and Colonization. At the end of 1991, the state expanded the Cuyabeno Wildlife Reserve to encompass all of Zábalo's territory. Large-scale oil production and uncontrolled colonization in the northern and western parts of the reserve had compromised much of

Looking downriver from Zábalo at dawn (Photo by Michael Cepek)

the area's ecological integrity. Under pressure from diverse social sectors, the government enlarged the protected area from 254,670 to 655,781 hectares.

At first, the reserve expansion angered the people of Zábalo because it meant that they could not secure a land title. Eventually, however, they changed their position when state and private actors interested in keeping oil out of the lower Aguarico helped them to form a residence agreement with the state agency that would become the Ministry of Environment. In July of 1992, after a series of Quito meetings between state officials and Cofán leaders, the "community"[5] of Zábalo became the legal possessor and manager of 82,000 hectares of territory. In 1999, Zábalo's landholdings increased by 55,000 hectares after subsequent Cofán petitions. An additional extension of 3,000 hectares in 2004 put a total of 140,000 hectares in the hands of the Zábalo Cofán.

My other main field site was the Quito Cofán Center (QCC). The QCC is a residential complex in northern Quito that houses the office of the FSC, the urban residence of the Borman family, and other buildings that Cofán students and visitors use as they pursue educational, political, and medical activities in the city. I spent a full year living at the QCC with Cofán leaders and students, including Randy Borman and his family. In addition, I accompanied Cofán people to the locations that they frequented across the city: schools, government ministries, NGO offices, stores, parks, and friends' homes. In many ways, the QCC exhibits practical and social similarities to life in Zábalo and other Cofán villages, making it a peculiar kind of Cofán community. Although the number of Cofán people residing at the QCC can swell to as many as one hundred individuals when there is an important political meeting, the average number during my stay was nine.

My main period of dissertation fieldwork took place from January of 2001 to December of 2002, broken by three month-long breaks during which I returned to Chicago to rest and analyze data. I lived at the QCC from January to June of 2001; in Zábalo from July to December of 2001; in Zábalo from February to June of 2002; and at the QCC from July to December of 2002. I made three shorter visits to Zábalo while based at the QCC in 2001 and 2002. Furthermore, I have returned to Zábalo, the QCC, and other Cofán communities multiple times since the end of my main fieldwork period: November and December of 2003; March, April, June, and December of 2004; April and May of 2005; July and August of 2006; May and June of 2007; June and July of 2009; July of 2010; and May and June of 2011.

In 2003 and 2005, I hosted Cofán people who traveled to Chicago to work with scientists at the Field Museum of Natural History. In 2008, I collaborated closely with three Cofán people who went to the museum to work on "The Cofán Historical Mapping Project," a project I had been pursuing with them since 2007. In 2003, I visited a Cofán student who was attending school in Seattle, Washington. In August and September of 2010, the student and his wife lived with me in San Antonio, Texas, as I tried to help them pursue university educations in the United States. The people of Zábalo realize that I have made a lifelong commitment to working with them. I intend to make good on my promise to return to their community at least once a year, as long as my body and my wallet permit.

As Zábalo's leader, Borman placed only two conditions on my research: that I learn A'ingae and that I live in the community long enough to understand the cultural dynamics of Cofán political experiments. As material reciprocation, I gave a large portion of my research funds to the FSC, and I provided the people of Zábalo with a satellite telephone and periodic medical aid. I also paid numerous Zábalo individuals as hosts and research assistants, which was a small but appreciated form of help at a time when virtually none of them had a steady income. Furthermore, I have a formal agreement to donate half of any royalties from *A Future for Amazonia* to the community of Zábalo and the other half to the FSC.

Randy Borman continues to be one of my main Cofán collaborators. He is the most important actor in contemporary Cofán politics, and he played an essential role in the creation of all the projects that I analyze. Recognizing Borman's skills, the Cofán have elected him to numerous positions: Vice President of Doreno, President of Zábalo, and President, Director of Territory, and Director of Ecotourism and Commerce for the Cofán ethnic federation. As the second decade of the millennium begins, he is the model for Cofán people's ideal leader. The residents of Zábalo and other communities are sending some of their children to outside schools with the hope that they will reproduce Borman's capacities, thereby making his contributions a more permanent feature of their political landscape.

My investigation depended on Borman's knowledge of historical detail, sociocultural nuance, and political circumstance. He studied anthropology in college, and he is an insightful ethnographer of Cofán society. In subsequent chapters, I include Borman's verbatim statements on virtually every aspect of the Cofán experiment. His reflections highlight some of the more complicated dynamics of the contemporary Cofán situation. Although he does not agree with everything that I write—and although he makes no attempt to "censor" my analyses, for which I alone am responsible—my learning profited from our many hours of discussion in Quito and Zábalo.

At times, my relationship with Borman can feel claustrophobic. I write about and alongside a "native" who can refute each of my assertions, not only to my face, but to anyone he chooses, as he has easy access to telephones, computers, and media outlets. In the end, however, the constraint is useful. My awareness that Borman can check me keeps me from making absurd claims. I give him copies of every article that I write, and he has read my doctoral dissertation in its entirety. He always offers criticisms, but he has never asked for changes. As far as I can tell, he believes that there is quite a bit of "truth" in what I write.

I have worked hard to avoid the slightest hint of hagiography, but my respect for Borman and other participants in the Cofán experiment is evident on every page of this book. Readers should attribute the positive tone of my account, however, to the objective conclusions of my analysis. Although I aimed for transparency and openness, I never would have entered a situation in which the subjects of my investigation demanded absolute control over its course. Thankfully, Borman and other Cofán people gave me the support and the freedom that I needed.

CHAPTER OVERVIEW

The ethnographic chapters of *A Future for Amazonia* are organized in two main sections. Part I deploys concepts essential to the study of political mobilization in an analysis of Cofán culture, social structure, and critical consciousness. Part II investigates the institutional forms that Cofán people negotiate and transform as they struggle to create a secure position for themselves in a world that appears to be bent on their destruction.

Part I begins with Chapter 1 (Agency: The Emergence of an Intercultural Leader), which offers an in-depth portrait of Randy Borman. I focus on the ways in which Borman's intercultural experience allows him to construct a creative, compelling, and workable vision for Cofán empowerment. Chapter 2 (Identity: Collectivity and Difference) examines the interethnic dynamics of Cofán history and sociality. I describe the substantive yet open nature of Cofán identity, which provides Cofán people with the interpretive means for including Borman in their daily lives and their political projects. In Chapter 3 (Value: The Dilemma of Being Cofán), I analyze Cofán social structure and value discourse. I explore the dual nature of Cofán social logic, according to which individuals pursue a desirable existence for themselves by delegating the negotiation of violent difference to such beings as shamans and chiefs. Paradoxically, these figures can only sustain Cofán life by participating in the forces that threaten it, and they thereby develop a deep ethnic ambiguity. From this perspective, Borman's emergence as an intercultural leader follows an authentically Cofán script.

Part II begins with Chapter 4 (The NGO: Institutionalizing Activism), which investigates the work of the FSC. I explore the political benefits and cultural challenges of NGO-based activism by analyzing Cofán experiences of such practices as grant proposal formalization, bureaucratic interaction with ethnic others, and "transparent" accounting procedures. In Chapter 5 (The Forest: Collaborating with Science and Conservation), I move from Quito to Zábalo to describe how the idea and practice of "conservation" transforms subtle dynamics of everyday life and expansive visions of global involvement. The chapter devotes special attention to Cofán participation in the scientific conservation projects supported by the Field Museum of Natural History, which works with Cofán people as they construct a new model for North-South collaboration. Part II ends with Chapter 6 (The School in the City: Producing the Cofán of the Future), which approaches formalized schooling projects in Quito as innovative attempts to reproduce Borman's powers in the next generation of Cofán leaders. I examine Cofán educational experiences against the background of urban life, whose pains and pleasures provide Cofán students with a new perspective on the personal and political value of an Amazonian existence.

The book concludes with a commentary on the practical and policy-related implications of Cofán experiments with indigenous and environmental politics. I end with a final reflection on the *longue durée* structure of Cofán political action, which is returning to the early militancy of the Conquest era after centuries of more passive relations with colonial and nation-state powers.

I AN INDIVIDUAL AND A PEOPLE

1   *Agency*

THE EMERGENCE OF AN INTERCULTURAL LEADER

My most definitive memory of Randy Borman involved a calamitous event. On November 3, 2002, the volcano Reventador, located at the southwestern edge of Cofán territory, violently erupted. I had taken a day trip with a friend to the town of Papallacta, which sits atop the Andes on the road to Reventador at a traditional passage point into the eastern lowlands. Early in the afternoon, we noticed small flecks of ash falling from the sky. Within hours, the sun was hidden behind a dark cloud. A thick gray sludge quickly covered the ground, and we began to fear for our safety. After trying vainly to wave down a bus on the nearly deserted highway to Quito, we found a family who was willing to drive us in their SUV. The ride was terrifying. Visibility was near zero as we slowly snaked along the road's perilous drop-offs.

Relieved, I arrived at the Quito Cofán Center in the early evening. When I walked into the FSC office, however, I was shocked to see that preparations for an expedition were under way. Instead of hunkering down in his Quito headquarters, Borman was trying to figure out how he could sneak past police and military checkpoints to watch rivers of lava flow from Reventador's core. In A'ingae, he said that Reventador was not just any volcano, but "ours." Cofán myths contain numerous references to the malevolent powers of volcanoes. Borman was determined to see with his own eyes what Cofán people had not witnessed for centuries.

On that afternoon, I imagine that Borman was the only Ecuadorian committed to reaching, rather than escaping, Reventador. With the aid of his brothers, he forged press passes. He concocted a story about being on assignment from CNN and the National Geographic Channel. With his family and a couple of FSC coworkers, he loaded his pickup truck with water, smoked game, and other provisions. Trying not to sound cowardly, I asked Borman if the trip was a good idea. With a smile and a wink, he replied that he could not think of a better way to die than to be engulfed in Reventador's flames. Months earlier, I had already learned that Borman did not scare easily. Yet seeing that excited glimmer made me wonder whether he was reckless and naïve. Two days later, however, he proved me wrong when he and his family returned with marvelous tales. They snuck their way to an overlook close to Reventador, where they watched the show.

Borman is not a traditional indigenous leader. Nor is he a conservative child of evangelical missionaries—let alone a colonial gringo on a fantasy flight from modernity. He is at home in many roles and situations, which he creatively, and sometimes mischievously, combines in his private and public personas. More than any other living Cofán person, Borman is the product of a multiplicity of cultural contexts: Cofán communities, lowland missionary bases, evangelical high schools, blue-collar American workplaces, cosmopolitan urban centers, and universities in both Ecuador and the United States.

Borman's movement across an array of locations reinforces the intercultural competence that enables him to act creatively in local, national, and international arenas. His creativity is an essential element of his political agency, which I define as the ability to distance oneself from the structures that compose one's life in order to reflect on and transform them. In a sense, Borman's work as a political agent depends on his imagination, which I understand in Paul Ricoeur's terms (1994). For Ricoeur, practical imagination involves three connected activities: first, the construction of metaphorical linkages between distinct fields of meaning and experience; second, the creation of novel images and ideas out of the conceptual articulations; and third, the artistic, practical, and political production of the envisioned objects and actions.

Borman believes that his success as a Cofán leader is related to important features of his biography. In this chapter, I take his lead and consider interviews, manuscripts, and firsthand observations of his life in Quito and Zábalo. Borman depends on Cofán, North American, and national Ecuadorian cultural repertoires. Nevertheless, he possesses a uniquely imaginative personality, which enables his work as a key architect of Cofán politics.

BIOGRAPHICAL SKETCH

Born in 1955 at a lowland missionary hospital in Shell-Mera, Ecuador, Randy Borman spent the majority of the first five years of his life at Doreno. He grew up with Cofán age-mates, speaking English with his parents and A'ingae with other villagers. At the age of five, Borman entered the one-room elementary school at the Summer Institute of Linguistics (SIL) base camp at Limoncocha, near the Napo River. His parents left him at Limoncocha for weeks and even months at a time. During vacations, however, he returned to Doreno with his parents and eventually his three younger siblings.

Borman spent his fourth-grade year at a school in Pasadena, California, when his parents were on furlough. He returned there for a semester of ninth grade. For the remainder of his high school years, he attended the missionary-affiliated Alliance Academy in Quito, where he was again apart from his parents. During holidays and summer vacations, he returned to Doreno and Limoncocha. On many weekends, he hitchhiked to hunt in both the highlands and the lowlands, alone or with friends.

After he graduated from the Alliance Academy in 1973, Borman spent the summer in Doreno. In the fall, he moved to Geneva, Illinois, to work in a furniture factory for half a year. He saved money to enter Michigan State University, which occurred the following January. After three difficult quarters in Michigan, Borman returned to Ecuador. He stayed at Limoncocha and Doreno before returning to Illinois in December of 1974 to work at a nursery school and take classes at Waubonsee Community College, where he stayed for a year. After Waubonsee, Borman returned to Ecuador. He spent time in the Amazonian region and then moved to Quito once more to take classes at the Universidad Católica. He studied at the university for only a semester, without earning a degree.

After college, Borman wandered through Ecuador. He lived for a short period on a parcel of land that he acquired through a government colonization program along the Aguarico River at Chiritza, where he attempted experiments with renewable forest products. He also stayed at Doreno and Limoncocha. In 1977, a missionary organization offered him a job in Guatemala to help rebuild after the major earthquake. He stayed for four months, returning by bus through war-torn Central America. With the income, Borman bought his first outboard canoe motor. He then began working with tourists who traveled to eastern Ecuador to experience the Amazonian environment.

In 1978, and at the age of twenty-two, Borman committed himself to a more or less constant residence with Cofán people. He lived as an active community member at Doreno, where he served multiple terms as an elected officer and helped secure the community's land title. Following downriver trips with Cofán friends and tourists in the late 1970s and early 1980s, he worked to form and lead the settlement of Zábalo. In 1987, he married Amelia Quenamá, a Cofán woman from Doreno. He spent the following years in Zábalo, partially shifting to Quito after his three children entered the Alliance Academy in the mid-1990s. From then until the present (2011), Borman has resided mainly in the capital. When he is not traveling to overseas meetings or coordinating projects in Cofán communities, he lives at the QCC with his family. During my main fieldwork years of 2001 and 2002, he spent less than half of each year in Zábalo.

Borman attributes much of his personality to his early upbringing at Limoncocha. According to him, life at Limoncocha was not doctrinally dogmatic. Nor was it a strange replication of conservative, rural America. He remembers the majority of missionary-linguists as "very open-minded, constantly searching for new ways of doing things, new information, new ideas." Residents downplayed the sectarianism that characterized relations between evangelical currents in the United States. People were focused on dealing with the major adjustments they had to make in preparation for fieldwork. At Limoncocha, Borman recognized the presence of classic American values: tolerance, cooperation, self-sufficiency, and a "jack-of-all-trades" openness to developing any skill that was needed for a physically and psychologically demanding existence.

Outside of school, Borman had a lot of free time. He explored the area surrounding the base. He wandered through the forest, fished in the lake, and hunted with a Cofán blowgun and an American pellet gun. On most of his trips, he was alone. "I was definitely more geared to the indigenous side of things than any of the other missionary kids. I was far more involved in hunting and fishing and all of that sort of stuff." It was at Limoncocha that Borman's twin loves—hunting and natural science—solidified. His early Cofán experiences prepared him for the former, and employment as an "interested gofer boy" for visiting field biologists introduced him to the latter. Borman accompanied scientists on their trips, during which he gathered specimens, took notes, and made collections.

Thanks to donations, the base at Limoncocha had a sizable library. Teachers encouraged the children to make use of it for research papers and independent reading. Borman recalls that when he was not hunting or fishing, he was reading a book. Adventure stories, historical accounts, natural science, and science fiction were his favorites. While on break from Doreno or doing linguistic analysis, Borman's mother returned to Limoncocha and saw that her son was "either reading a book or out in the jungle, one of the two." In an unpublished autobiographical draft, Borman reflects on the desire for adventure and heroism that he learned from reading:

> Reading opened up a horizon that rivaled the view of the jungle from the airplane. Here were sweeping stories of disciplined armies pouring across natural and human boundaries to change the course of history. Here were young people who had suddenly been thrust into impossible situations and survived to defeat the odds and become heroes. Here were hunters who made even the older Cofáns facing the jaguar with wooden spears look rather shabby. This new world had mammoths, tigers, and

dinosaurs. The plains of Kenya, the forests of India, the vast bison herds of the western United States, all made my wild pigs and caimans look sort of tame. . . . What was getting chased into a boat by a lousy little caiman, or growing up in an Indian village on the headwaters of the Amazon, to these glowing adventures? I began to split my time between my real excursions and my excursions of the mind. My mom tried in vain to console me by pointing out how exciting my life was compared to most kids', but I was deaf to her admonitions.

By the time Borman entered high school at the Alliance Academy, he felt that his development as an autodidact and his physical conditioning via hunting and forest exploration gave him an easily recognized adult aura. Dorm parents allowed him a tremendous amount of freedom. He soon attracted a group of students who sometimes followed him on his weekend hunting trips in the Andean valleys surrounding Quito. Nevertheless, the more urban missionary children recognized him as one of the "Limoncocha kids." SIL children were mature and responsible, but they were strangely socialized: "We stank. We didn't know what to wear. We were using the wrong color combinations."

Borman was never very interested in high school. He earned low grades and proved himself a recalcitrant rebel. He tested adult estimations of his sensible stubbornness by repeatedly growing a moustache and being punished for it. He entered actively into the schism between the administration and the students during a period when the countercultural movements of the United States stimulated the political imaginations of the missionary youth. The students read Marx, flirted with Liberation Theology, and debated Ecuador's poverty. Borman reflects:

> My class was very socially conscious. I guess you'd call it "antiestablish-
> ment." There was a lot of idealism. "Well, I can't see buying into this whole
> establishment system." Of course, I was part of the discussions. But I think I
> was one of the few people who just assumed that we were talking seriously
> about our futures, rather than a stage that we were going through.

During his senior year, Borman's class elected him the "mediator" between the student body, the faculty, and the administration. The position was both a profound leadership experience and deeply alienating. The conservative opinions and political apathy of the adults disillusioned him. The immaturity of the students' ethical explorations did the same. Borman sees the political and ethical soul-searching of his late high school period as an integral part of his unconventional life trajectory.

Although Borman had been spending summer vacations among Cofán peers in Doreno, it was not until he graduated from the Alliance Academy in 1973 that he made a more in-depth return to Cofán society, settling at Doreno for three months. Although he missed his Cofán friends, his primary desire was for the forest:

> I was absolutely in love with the forest and the hunting and the animals and the whole thing. That was my primary focus. Cofán culture was a convenient way to get there. And in the process I was developing really solid relations and really solid friendships. I don't think my interest in the forest would've been there without the Cofán relation. It was an assumed part. But I was way out beyond my average peer, Cofán peer as well as gringo peer, when it came to my fascination with hunting and the forest.

Nevertheless, Borman's formal education, first at Limoncocha and then in Quito, kept him away from the forest and the hunt. He missed out on the practical education of his Cofán friends, who accompanied older relatives on extended treks. Borman remembers his return to Doreno as fairly stressful. Even today, he blushes when he recounts how Cofán people laughed when he did not follow the proper protocol for butchering game. In order to transcend the disparities, Borman turned to his indirect, science-based knowledge:

> I had a highly developed dual sensitivity to the environment. One, from my reading and my scientific background. Two, from my Cofán background. I was constantly combining the two to upgrade my abilities. I couldn't keep up with my Cofán peers on the basis of actual daily living experience because I was stuck in school for such long periods of time. But I could keep up with them in my actual field abilities by taking what I was learning in the Western world and combining it with my Cofán background. I'd read every bit of information I could find about an animal that I was interested in hunting. All the old explorers' accounts, the scientific accounts, and everything else—to tip the scales a little bit. That's why when I first really came into the Cofán community as an adult, I was a much better hunter than most of the people. And much more fearless. I would go out three or four nights alone into the forest. That sort of thing didn't even faze me.

Borman's summer in Doreno ended with his move to Geneva, Illinois, in the fall of 1973. The shift marked the beginning of his two post–high school years in the United States: five months working in a furniture factory and living with his grandparents, seven months studying at Michigan State University, and a full year working in a nursery while taking classes at

Waubonsee Community College. (The summer preceding Borman's entrance to Waubonsee, he returned to Ecuador to conduct an independent study on caimans at Limoncocha, for which he received credit at Michigan State, and he also spent time at Doreno.)

In the United States, Borman discovered that he was missing many essential "living skills." He did not know how to write a check, drive a car, or talk about sports. The experience helped him understand a key aspect of his persona. On the one hand, he was a white man who could speak fluent English without an accent. On the other hand, his worldview reflected his immersion in national Ecuadorian, North American missionary, and indigenous Cofán cultures. Borman feels that his evangelical background and private sojourns had "insulated" him from moral and psychological crises. Nevertheless, he felt like an outsider. He watched Americans and reflected on their culture in a near-ethnographic style:

> I was very much the anthropologist, observing the quaint rituals of this extremely interesting cultural phenomenon but without any real cultural point of contact myself. And unlike most anthropologists, I already had the language down. I already had the looks down. I was able to slide in in a way that most anthropologists would love to be able to do. Yet I was very much in the observer mode. My whole time in the States I was observing.

Listening to Borman relate this phase of his life, it appears that he simply did not feel the need to "fit in." He had many casual friendships, but he was always one step removed. He describes the difficulty as one of having no "real point of connection" with the United States. A well-founded conversation would have required the presentation of an immense amount of background information. More often than not, Borman left his own story aside. He entered social life with only half his heart: "I never felt part of anything." He did volunteer as a counselor with a Christian youth group in northern Illinois. Nevertheless, he spent more time riding his bicycle, hunting with a pellet gun, picking edible plants, and collecting fresh roadkill to cook for himself.

Borman's return to Doreno in the fall of 1974 intensified his sense of distance from the United States as a residential and cultural location. The Michigan experience made him realize that the United States was not his "home." Returning to Ecuador, he expected a renewed sense of belonging. Most of his previous non-Cofán social circle, however, had left. Many of his friends from the Alliance Academy were studying in the United States. His parents were also there for a year. And many of the missionaries at Limoncocha had moved to new locations.

When Borman finally made it to Doreno, he found a community that had stayed the same. It felt like an island of familiarity in a sea of change and difference. At the same time, the community appeared to be at risk with the encroachment of the oil industry and uncontrolled colonization, both of which had grown in scale since the beginning of commercial extraction in 1972. Together, the sense of return and the sense of crisis anchored Borman in the Cofán world:

> I came back and found that, yes, this was my home. I went to Doreno, and all of a sudden I was home again. After essentially a year away—and in a broader sense, ever since high school began, away—there was this group of people whom I knew, and who knew me. There was a social stability. There was a social infrastructure. That was extremely important to me. And then when I got there, I found that all of the things that I was focusing on as home—my whole community structure, my whole social structure, my whole physical world structure—were being eroded rapidly by the Western influx. Alarm bells started going off in my head, and I said to myself, "If I'm going to wind up with anything that is meaningful to me, I'm going to have to help out."

Recognizing the personal importance and precarious situation of Doreno led to the birth of Borman's political consciousness. He recognized the Cofán way of life as a supreme value in his life, and he realized that its existence depended on a stable land base. He mobilized community members to make boundary trails. He also worked through SIL contacts to pressure President Rodrigo Lara and the Ecuadorian Institute of Agrarian Reform and Colonization to recognize the Cofán land claim. Borman was still a teenager, however, and he had to negotiate his confusing reception as a Cofán spokesperson:

> I was still very, very conscious outside of the village of being gringo. I would talk about "we of Doreno," but I wouldn't talk about "we Cofáns" to outsiders. It had nothing to do with the way I was feeling inside the community. But I felt like I couldn't take a primary leadership role. I had to work through other young Doreno people. [Were you ever questioned by political officials?—M. Cepek] I assumed it was a problem. I was uncomfortable with it. I had to explain many, many times what I was doing there.

When Borman returned from Waubonsee at the end of 1975, he moved into his parents' Quito home. It was the first time his family had lived together in more than six years. He bought a cheap motorcycle and spent weekends hunting in the lowlands, in Doreno and elsewhere. (His mother joked that she was the only woman in Quito who had to deal with spider monkeys being cooked in her kitchen.) Beginning in January of 1976, Borman took classes at the Universidad Católica. He found the national education system to be "unbelievably trite and ridiculous." He criticized his professors and peers for downplaying the importance of "tangible evidence." Moreover, he recognized an overly politicized drive not to accept any knowledge perceived as originating from First World countries. In addition, he was dismayed with the cheating and class- and kin-based networking that dominated Ecuador's universities. After a semester at Católica, he dropped out. He was disillusioned with the Ecuadorian education system and "national culture" in general, at least in its middle- and upper-class variants.

During and after his studies at Católica, Borman began to work at a new location: a parcel of land that he acquired under a government colonization program at Chiritza, a still-remote location downriver from Doreno and upriver from what would become Zábalo. Chiritza was Borman's first "experimental" project. In the small rainforest territory, he combined his knowledge of ecology and tropical agriculture to "alter the forest ecology so that you have a predominance of valuable products" without clearing the bulk of the forest. He weeded out some plant species while encouraging the growth of certain palms and hardwoods. The project never took off, but it motivated Borman to think on a larger scale. He envisioned something along the lines of the SIL base camp at Limoncocha—with its diversity, technology, and cooperation—on a much larger reserve of forest inhabited and worked by Cofán residents.

Borman abandoned the experiment after a few years of discontinuous work, but it represented an important stage in the development of his political aspirations. For the first time, he conceptualized the creation of a socio-ecological form as a practical, political project. And, even more importantly, he attempted to realize the project. Eventually, and in a much closer relation with other Cofán people, Borman returned to the experimental impulse by forming Zábalo as a unique kind of indigenous Amazonian community.

Short on money and open to adventure, Borman accepted an offer from a missionary contact to travel to Guatemala in 1977 to help rebuild after the major earthquake. There, he learned important skills, including construction techniques. He went back to Ecuador after four months with over a thousand dollars. After his return, in early 1978, Borman made a more or less total shift to life with Cofán people. With the money, he bought his first outboard motor, which is the most expensive and desired commodity in contemporary Cofán society. With the help of an extended kin group, he built a fifteen-meter dugout canoe. The people who aided him were from the Criollo family line, themselves migrants to the Aguarico region from the San Miguel dialect group two generations earlier. With the building of the canoe and the purchase of the motor, Borman's affiliation with the Criollo family strengthened. Today, they form much of Zábalo's population.

With the canoe and motor, Borman also began an important Cofán tradition: ecotourism. At Limoncocha, visiting missionaries and scientists employed Borman to drive motorized canoes. Over time, he learned to provide a running commentary on the sights and sounds that confronted the newcomers. In 1978, an American man passed through Limoncocha wanting to see the famous Amazonian dolphins. People told him about Borman. With the new canoe, Borman and two other Cofán men brought the visitor down the Aguarico and into the Cuyabeno lakes for a trip that lasted nearly a month. The visitor covered all the expenses. In what became an early pattern, Borman and the other Cofán understood "tourism" as all-expenses-paid hunting trips to areas where game was plentiful.

With the influx of colonists and the felling of the forest around Doreno, the mounds of smoked meat that Borman brought to the community were deeply appreciated. Eventually, more and more people from Doreno became involved. After visitors helped create contacts with agencies in the United States, Borman learned that tourism revenue could fulfill a Cofán community's basic needs for commodities, which were still quite limited. As the wildlife surrounding Doreno disappeared in the early 1980s, Borman and his coworkers took tourists farther and farther downriver. Over a decade, the trips gradually led to the settlement at Zábalo. A core group of Cofán working with Borman began to build huts and plant gardens at the new site, which was more than one hundred miles from Lago Agrio.

Borman's total immersion in Doreno life in 1978 occurred at a precipitous moment. The state had just legalized the community's 9,451-hectare territory. Consequently, Cofán people had to create a governing political structure, which was a prerequisite under Ecuadorian law for Doreno to have full *comuna* status. The Cofán needed to organize a governing body modeled on a Western system, with a president, vice president, treasurer, secretary, and *síndico* (legal advisor/representative). Doreno residents elected Borman *síndico* first and later vice president. He felt that his political reception in the community was "totally natural." No Cofán questioned his identity. Nevertheless, becoming a legal Cofán official made him a mediator with the outside world, a role with which few other residents felt comfortable.

From the late 1970s to the early 1980s, Borman and the Cofán people who worked with him on tourism were leading a mobile, dual-residence existence. They moved back and forth between Doreno and the land that would become Zábalo. By the mid-1980s, Borman and two extended families built houses and fields at Zábalo. Another extended family followed them at the end of the 1980s. After Borman married Amelia Quenamá, many of his in-laws also moved to Zábalo. It was a period of solidification for the Zábalo Cofán, especially during a series of tense land conflicts with Napo Runa communities that bordered them both upriver and downriver.

By the time Zábalo became a stable community, Borman was aware of his leadership position. People's willingness to follow him to Zábalo was based on many precedents: his leadership of the effort to secure Doreno's land title in the mid-1970s, his elected positions at Doreno, his organization of tourism operations, and his handling of negotiations with both neighboring indigenous peoples and the government to secure a land base at Zábalo. Ex post facto, Borman considers the development of Zábalo as the success of what he first attempted at Chiritza. The missing ingredient, however, was Cofán perceptions of him as a capable leader:

> What I had tried to establish at Chiritza was basically what I wound up with at Zábalo. But I didn't have the status, the pull, to be able to coalesce people around me. I was still a young whippersnapper, an unknown quantity that nobody wanted to trust and actually follow. To go down there, work for a while, and pick up some meat, sure. But not really to go down there and stay to live. I was building what is now looked at as community-based ecotourism. But that was evolving. I was very aware of the fact that Doreno was too small. We didn't have enough wilderness to support our group. That was partly a historical recognition, because I knew what had happened in North America and South America to indigenous groups in the face of colonization. So I understood what was happening better than anybody else.

THE COORDINATES OF IDENTITY

When Borman talks about his biography, it is clear that he did not feel particularly conflicted about his identity before his late adolescent years. As a child, he sensed his distance from many social contexts. Nevertheless, he did not feel the need to choose membership in one group over another. After he graduated from high school, however, the question of what kind of life he would lead posed itself. As he explained, "My question, as I got out of high school, was 'How can I personally continue to live in these "pristine" environments, practicing this way of life that I love?'"

The missionary community made Borman feel much more anxious about his identity than the Cofán. Cofán identity is open and malleable, and the question of divergent "careers" is practically nonexistent for a people with a flexible social structure and a generalized division of labor. In the missionary community, however, there was a strong pressure for "missionary kids" to return to the United States. They had to show their friends and families that they could succeed as American citizens before contemplating a return to missionary work. To remain in the country of one's parents' assignment was the "failure" of a "problem" or an "oddball" child. Even worse, most missionaries understood marrying a local and living in a small-scale community as the sad story of someone who could not handle reimmersion into American society.

In the environment of anxiety, Borman began to use a vocational metaphor to conceptualize his life choice. In high school and university and church, he had to formulate a response to the questions, "What do you do?" or "What will you do?" Borman took the query as an invitation to play the part of what Charles Taylor calls a "strong evaluator" (1985). In other words, he chose the kind of human being that he would become by deciding on the value of his values. In a paper that he delivered at a missionary-attended conference on "Third Culture Kids" in 1986, he writes:

> [T]he missionary and church community offered me three culturally acceptable alternatives: (1) return to live in the U.S., pursue my education, and become a professional; (2) live in the U.S., get a job, and settle down to a middle-class suburban existence, or (3) become a missionary or professional Christian worker in the traditional sense of the word. The third possibility was the only one that really made much use of all my years of training in living in different cultures and places. Yet, as a "native" rather than a "foreigner," I felt there were a lot of things I'd do differently from the way traditional missions were equipped to work. In this I received very little encouragement or guidance. After several years of "on-site training," if you will, I found myself deeply involved in the progress of my hometown of Dureno, and concerned intimately with the material and spiritual lives of people up and down the Aguarico River system. . . . I found that my original "family," as it were, of missionaries and church people had a very hard time accepting what I was doing. I heard "cop-out" more than once, and people would make references to "when he settles down" and "what do you plan to do with your life?" And, it wasn't until I began to make "American-style" money as a tour guide that I really felt that the pressure began to ease—somehow, even though a bit eccentric, I was a success and the missionary community no longer had to shake its head sadly and say, "Too bad, another mixed-up kid with problems."

Borman's acceptance as a full-time member of a Cofán community was never an issue. Neither he nor other Cofán people remember situations in which a resident of Doreno or Zábalo questioned his status. His only concern within Cofán society was whether he could prove himself in terms of a forest-based way of life. When outsiders ask how he understands his relation to Cofán society, however, he still lapses into the vocational metaphors that he developed in interactions with missionaries. For example, he told Mike Tidwell, author of *Amazon Stranger* (1996), that he thinks of himself as "operating an indigenous community."

Borman's willingness to play with the terms of his life rides on his hard-won comfort with outsiders' perceptions. After years of visits from entranced North Americans and Europeans, he is confident in their approval as well as his powers of self-explication. Borman loves to push his interlocutors in novel directions, forcing them to accept concepts or statements that at first appear strange, contradictory, or objectionable. He relishes the idea of the uniqueness of his life, and he is most clearly able to express his particularity in interactions with Westerners, whether tourists, journalists, or anthropologists.

Below the highest level of identity as a life choice or "job description," Borman employs a number of terms to describe his embodiment of multiple cultural resources. He talks of both Cofán and national Ecuadorian "cultures" and "worlds." He often places "American" or "North American" as the third position in the triad. If you ask him to list his "identities," he will tell you that he thinks of himself first as Cofán, second as Ecuadorian, and third as American. Usually, though, he talks about his "biculturality." He focuses on the first and third terms as cultures in which he deeply participates. The second is more a matter of "citizenship," although it is still important in terms of his allegiances and affiliations. (Borman does have dual American and Ecuadorian citizenship.)

Although he sometimes differentiates Cofán identities according to the beliefs and practices of communities (Doreno vs. Zábalo vs. Dovuno, etc.) or dialect groups (Aguarico vs. San Miguel), Borman's most complex identity is "American." He is clearly aware of the realities of race, class, and gender. Nevertheless, he typically organizes the American cultural category into a divide between "(evangelical) Christian" and "Western scientific/materialist" sides. His own identifications move back and forth between the poles, which he has combined in a hybrid position that no longer troubles him, as it did during his adolescence.

Borman describes his "biculturality" as having his "feet in two cultures," being "in a unique position between two cultures," possessing different "hats"

Randy Borman (right) at a meeting in the community of Chandia Na'e (Photo by Michael Cepek)

that he can wear in different contexts (thereby acting as a "chameleon"), and having a Cofán "half" and an "American" or "Western" half. He prefaces some statements with either "from a Western background" or "from a Cofán background." The assertions that follow depend on a certain cultural and epistemological framing. For example, during a conversation with a visiting scientist about the disappearance of white-lipped peccary herds from Zábalo's forests in the late 1990s, Borman suggested that "from a Western background," he would consider the interplay of such factors as hunting pressure, random herd movements, and disease. "From a Cofán background," in contrast, he would point to the strange manner in which jaguars were killing, but not eating, the peccaries—a sign of hostile interventions from enemy shamans. Science and Cofán cosmology supply Borman with constant stores of explanatory devices. He can combine the two to come up with a more complete account of any event, which he proposes with an eye to the immediate audience.

At an even more implicit level, Borman describes his "halves" as "senses" or "sensibilities." Sometimes, he uses the terms *attitudes*, *outlooks*, *mentalities*, and *values*. The terms are common in Borman's discussions of his appreciation for hunting and the Amazonian environment. Describing his early attitudes, he writes in an unpublished manuscript:

On one hand, I was a Cofán in my attitudes, skills, and outlook. This "Cofán-ness" included my attitudes toward the forest and its animals. . . . [T]hese attitudes were basically pragmatic. My natural aptitudes and skills led me to become an expert hunter and woodsman, and it was via my hunting skills that I first proved myself as a Cofán. And at first, attitudes toward the animals I hunted were very basic ones of how can I kill the greatest number of animals the easiest—a pragmatic set of values that had served the Cofáns of generations well.

Sometimes, Borman describes his biculturality as only partially under his control. He talks about his negotiation of competing senses as a matter of long-term learning. Only gradually and with much stumbling did Borman succeed in melding his Cofán sensibilities with his scientific and conservationist understandings. He swears that it was as hard for him as it was for any other Cofán to forego shooting a woolly monkey or a macaw upon encountering it. With gun in hand, the impulses are difficult to deny in a Cofán, forest context. With the solidification of tourism and the creation of conservation strategies, however, he had to learn how to let a new set of sensibilities take over.

In short, Borman's internal sense of identity is organized in a nested structure. At the most encompassing level is the concept of life choice, written in the idiom of vocation. Borman partially articulated his commitment as an explanation to the North American missionary community, whose language of soul-searching and free will he shared. As a Cofán political leader, Borman's job description implies an identification of his biculturality, or his unique position between two (and sometimes three) cultures, with their own internal complexities. At the most implicit, embodied level, each of the cultures forms a set of senses and sensibilities, which Borman attempts to integrate in a productive fashion.

In the mid-1980s, after the formation of Zábalo and his marriage to Quenamá, Borman's internal sense of identity appeared to be set. Nevertheless, his most active days as a Cofán spokesperson were yet to come. In the early 1990s, Borman led the people of Zábalo on a campaign against the oil industry. From 1994 to 1996, he served as the elected president of the Cofán ethnic federation. His political representation put him into contact with the global media, transnational activist circuits, government officials, and the national indigenous movement.

Borman has always felt somewhat strange about his reception by various sectors of Ecuador's indigenous and nonindigenous population. Most North Americans and Europeans understand him as an interesting and supportable if totally unique cultural phenomenon. Highland indigenous leaders, however, have questioned him because of his gringo heritage. Lowland indigenous leaders, who have less experience with an explicitly race-based system of oppression, have been more open to his political position. Nonindigenous Ecuadorians only rarely question him. Borman assumes that his reception among mestizos is the result of his incitement of diametrically opposed instincts: to denigrate the forest Indian and to assume the superior education and power of the white North American. In his words:

> Among the Ecuadorian government people there's a confusion. Because here's a gringo who's automatically accepted into that upper level of society just because he's a gringo. But then to find out that this gringo, whom you have to accept as a social equal, is also an Indian—it twists their world thoroughly. It makes it very difficult to know how to relate. Because an Indian is something that you kick in the butt and offer him $1 for his potatoes. That's what an Indian is. Automatically dirty, automatically socially unacceptable, automatically a nonentity. To have that combined in one person throws government people.

Borman's most important source of support as a political representative comes from the Cofán people themselves. They have repeatedly elected him an officer at both the community and national level. In a recent bilingual FEINCE publication, *Ingita gi A'indeccu'fa* (*We Are Cofán*), there is a full-page picture of Borman directing a meeting. No words differentiate him from the other representatives who appear in the book. The message is clear: Borman is a Cofán leader, just like the others.

Borman became aware of the Cofán people's desire to position him as their national representative during the oil struggles of the early 1990s. On one trip, he accompanied three men from Zábalo to Quito as they gathered media attention and outside support. Although still uncomfortable donning Cofán regalia on the global stage, Borman acquiesced when his Cofán companions told him to put on his headdress, jaguar-tooth necklace, tunic, and face paint:

> When we were getting ready to go over to the Fundación Natura [a Quito NGO], Lorenzo, Atanasio, and Mauricio [Zábalo elders, who always wear traditional clothing] said, "You're not going in this suit. You're going dressed up as *a'i* [Cofán]. And we're not going like this if you're not." And that's when I told them, "Well, I feel kind of funny going in like this." And they said, "We're the ones who can say whether you have that right or not, not them. Don't worry about them. We're the ones who have the right of

decision." So I said, "Okay." That was the point where I really took on my role, my identity in the outside world as a Cofán leader.

Borman is more than a passive product of his immersion in multiple cultures. As a political agent, he engages in a constant stream of reflection on the social worlds that compose his life. Borman's agency profits from what I call his "structure of certainty": a system of conceptual coordinates that he uses to reimagine American missionary, national Ecuadorian, and indigenous Cofán cultural milieus. A coherent scheme provides Borman with a set of deep, highly theorized principles, which he uses to evaluate his models of and models for sociocultural reality. His structure of certainty depends on his multicultural biography in a highly distilled sense, as it consists of the propositions that he has identified through faith, experience, logic, and formal education.

Much of Borman's orientation to life and death comes from his connection to evangelical Christianity. While living with his family in Quito, I witnessed many indices of Christian belief in his everyday routines: prayers before meals; nightly "devotions," or religious teachings, with his children; and fairly regular attendance at the English Fellowship Church, an evangelical congregation popular among Quito's missionary community. In Zábalo, he sometimes leads Sunday Bible discussions. (More often than not, however, there are no services.) Borman is even a titled "minister," capable of performing marriages. It is important to note, however, that Borman laughingly told me that he received his license from a mail-order institution advertised in the back of a magazine.

Borman thinks of himself as living by the two most important commandments of Christianity: to love God, with all one's heart and resources, and to love one's neighbor, which includes both the "spiritual" and the "physical" person. The latter principle justifies his earthly political work. He prays regularly, seeks divine guidance, and actively identifies and elicits the "hand of God" in everyday life. While speaking to other evangelical Christians, he talks about God's "word" and "will." Often, he looks back at the success of certain actions and decides that they were "blessed." Early in my research, he advised me to be open to the possibility of God's influence in what I was studying. Familiarity with and openness to God is paramount in Borman's life. More than anything else, Borman considers God's presence to be the ultimate condition of all that he is and does.

Not surprisingly, however, Borman's evangelical Christianity is not typical. He has strong universalist tendencies. He believes that God can be present to peoples before the arrival of missionaries or the Bible. Moreover, he identifies Christian "maxims" and "attitudes" in cultures that do not claim to be Christian. In addition, he believes in shamanic efficacy, and he has undergone shamanic curing for a number of ailments. He is quick to explain, however, that he has an ambivalent opinion of shamanism, which he considers to be "self-centered" rather than "selfless." Nevertheless, Borman experimented with shamanism during his youth, and he has resorted to shaman-like actions in his adulthood.

Borman told me that during a conflict with an upriver Secoya community in the 1980s, he conducted a "jaguar curse" to remove newcomers from Cofán land. Shortly after his performance, Borman attests, a jaguar entered the territory and killed the pigs that the Secoya had brought without eating a single one. Borman believes that the jaguar's action was a direct response of God to his call. He sometimes narrates the event to North American Christians, who are usually troubled. Nevertheless, he follows the story with an argument explaining God's many uses of wildlife to perform his will, as described in the Bible. From his perspective, the ultimate point is whether people understand themselves as agents of God or whether they operate according to their own self-interest. If the former is the case, then he is open to the mediation of any number of natural or supernatural entities in the performance of God's will.

Borman's strong faith in science shapes his Christianity. He describes his worldview as consisting of three elements: "science" (the rational-empirical-scientific outlook, which is basically "agnostic"), "Cofán cosmology/shamanism" (the "indigenous side," which holds that the spiritual and the supernatural are constantly present in everyday life), and "American Christianity." He insists that Christianity is the mediating element because its vision of reality lies somewhere between the disenchanted world of science and the thoroughly supernatural world of Cofán cosmology.

Borman believes that science, Christianity, and his Cofán background are compatible. He claims that he has "seen" the efficacy of the spiritual, whether in his jaguar curse or in the ability of Doreno's old shaman to call herds of peccary to the village center, on command. He quotes the principle of "Ockham's razor." From his perspective, the best and most scientific explanation is the simplest, even if it involves firsthand experience of supernatural events. For Borman, "magic is real." He argues that there is no other explanation for events that he has seen with his own eyes.

Borman is also a believer in evolutionary biology and contemporary astrophysics. Even more than his relation to Cofán cosmology, his scientific understanding sets him apart from other evangelical Christians. His conceptualization of a universe that is billions of years old and amazingly complex in scale and structure reinforces his sense of the power and grandiosity of God. This "cosmology" makes Borman feel somewhat alienated from most contemporary Christians. He believes that they embrace an overly anthropomorphized conception of the divine because of their failure to imagine the kind of God who can control such an immense and impressive cosmos.

Borman's willingness to accept complexity, which amounts to an admission of human ignorance, allows him to reconcile his scientific outlook with belief in the afterlife. His openness to an afterlife, as well as his understanding of an all-powerful and eternal God, verge on an antihumanism. From his vantage point, the life span of an individual human being appears to be of miniscule significance in the grand scheme. For Borman, truth, reality, science, and God are of a piece. He realizes that his grounding in his structure of certainty distances him from the members of every cultural community in which he participates. Furthermore, his conscious orientation to the overarching structure allows him to undertake radical actions in earthly affairs, whose ultimate significance and attendant dangers appear reduced in their encompassing context.

Overall, Borman's understanding of his life in the frame of a spatially and temporally expansive universe—created and controlled by God, and described and revealed by science—provides him with a fatalistic and risky orientation to action. Borman has faith in his grasp of moral and epistemological absolutes, which allow him to act decisively on the basis of his own evaluations. And with little fear of the wrongness of his actions or the fragility of his life, he is particularly suited to think and act creatively—and to take substantial risks. Borman feels morally and empirically justified in his political pursuits, even when they appear dangerous, irrational, or counterintuitive to other people.

There is one more aspect to Borman's structure of certainty: his knowledge of world history. Borman is a voracious reader, and history is one of his favorite topics. His historical orientation allows him to understand current events and future possibilities as tokens of the social and natural types that are present around the world and through the ages. In an unpublished telling of "The Zábalo Story," he stresses how his historical studies differentiate him from other Cofán people:

> I was as poorly equipped for dealing with these outsiders [colonists and oil companies] as any of my peers—except for one thing. I loved and had deeply studied history. My background in history allowed me to see with far clearer eyes what was at stake here in this world, which I had claimed for my own. The Romans, as they pierced the forests of Gaul and Germany, had shattered forever the tribal systems that had been in place since the early

Bronze Age of Europe. The English and the French had done the same to the late Stone Age peoples of the forests of North America. The Spanish experience had been very different, as they came into contact with far more advanced and complex agricultural communities, but now, as their descendants marched into our Amazonia, the story and the system was once again in pattern. . . . I knew what the patterns were. I knew what had happened elsewhere. I knew what had happened in the U.S., both to the indigenous peoples and to many of the native species. I knew that if we did not have land, our culture was doomed.

In short, Borman's structure of certainty provides the moral and conceptual coordinates that he uses to reflect on his life and make his decisions. It consists of a system of knowable absolutes: an all-powerful but righteous and helpful God, who provides an eternal interpretive frame for all action; a universe revealed in its truth by science; and a store of possibilities for social and natural events that are manifested in the repetition of limited, historically apparent patterns. Borman's confident creativity is grounded in his conscious orientation to what he understands as clear and consistent parameters.

POLITICAL WILL

Borman's approach to existence is also profoundly literary. He grew up thinking of his life in relation to stories of danger, adventure, and heroics, which he gathered from the Bible, historical accounts, and books of travel and science fiction. In both the figurative and literal senses, Borman writes his life story. In Quito, he spends many days in front of his computer, composing proposals for government agencies and granting foundations and NGOs. He has published a number of more or less academic articles, which cover such topics as missionary influence and culture change (Borman 1996), ecotourism and community development (Borman 1999), and the early history of Cofán territory (Borman 2009).

In the late 1990s, Borman worked with a literary agent to begin writing an autobiography, which he never completed. In addition to the proposals, articles, and book manuscripts that might one day find their way into print, Borman writes short pieces for no apparent reason at all, except diversion or self-clarification. While living with him in Quito, he sometimes tossed a few pages to me at the dinner table. They contained his musings on the history of the Americas or relations between indigenous peoples and "macrocultures" or new principles for novel indigenous land-rights regimes in nation-state contexts.

Most of Borman's writings pertain to his life, whether personal or political. His constant narration is another aspect of his reflexivity. His life is very much an object for him, and he regularly describes it in written and oral forms. Government officials, tourists, anthropologists, and journalists have asked him numerous questions about his story. He is very conscious of the fact that he has a global persona produced in large part through media images, which he attempts to co-construct. While reviewing the first- and secondhand accounts of Borman's life and comparing them to my own research materials, I am struck by the consistency of his reflections. Borman's continuous line of self-understanding suggests that, in many ways, his reflexivity takes the form of an active scripting of his life.

Considering the features of Borman's script allows for a unique view into the basic motivations structuring his political will. Whereas I cannot reproduce in full his autobiographical writings, I wish to describe two episodes that he has put to type, if not to print. Neither is published, and neither is part of a completed manuscript. Therefore, as an author, Borman would not want us to interpret them as fully formed.

The first piece concerns Borman's portrayal of the moment before the Zábalo Cofán began their fight against the oil industry in the early 1990s. Borman writes the story in the genre of an adventure tale. He shows a sensibility that appears to derive from the rugged, self-sufficient ideals of mid-twentieth-century America. The content of the narrative is similar to the storyline of a Louis L'Amour or James Michener novel, with their portrayals of against-the-odds situations overflowing with possibilities for heroic action. Describing the mindset of the Cofán prior to the beginning of the conflict, Borman writes:

> Our village Bible studies for a year or so had been slowly working their way through the Old Testament. It had been easy for us to identify with the early Israelis as they lived their seminomadic lives in a land that, if not exactly a forest, certainly resembled it in many ways. . . . The pact between the Creator-God and a tiny and vulnerable people was not only a distant piece of history but an intriguing possibility for a community that was feeling very tiny and very vulnerable. Hence it was curiously appropriate that our studies that October were from chapters six and seven of the Book of Judges—the story of Gideon. Here was a blueprint for action—a tiny group of determined people with thoroughly unorthodox weapons who trusted in God for victory and charged forward against overwhelming odds. God sent the foe reeling away in defeat, never to return to bother the Israelis in that generation. But it took three hundred people who were willing to attack an army of thousands upon

thousands of veteran raiders with nothing but trumpets, clay pots, and torches. It also took a leader who could get those three hundred to do it. Actually, Gideon's strategy was excellent. Attacking a poorly organized army of several different language groups by night, after a great deal of propaganda, and giving the impression of a huge army—then not providing anybody for the enemy to rally against. The confusion must have been incredible, and I could quite believe the various language groups mistaking each other for the enemy and going at it in a thousand melees in the absence of any Israelis to fight. I could only hope I would be as good a strategist against our enemy. . . . [B]ecause it was now obvious that we were on our way toward carrying out the contingency plan we had talked about way back in those faraway days when we had first heard that there was the possibility of oil exploration in the region.

In the passage, Borman portrays himself as a leader of Cofán warriors who are about to embark on a movement of near-military proportions. The David-versus-Goliath scenario is also apparent. The premise relates to one of Borman's broader characteristics: his antihierarchy impulse. It is tempting to trace the sentiment to classic values of North American pioneer culture, according to which a small group must strike out on their own to fight against preordained inequalities, thereby ensuring their autonomy and survival.

Borman's political imagination focuses on overturning the structures of domination that place Cofán people at the lowest level of cultural, political, and economic power. He understands himself as an underdog in a fight against a much larger enemy, with the odds squarely against him and the people he represents. We could interpret the struggle in overtly Christian terms: fighting so that the last will be first, and the first last. Or, we could imagine him as one of the archetypical figures of history and literature, especially the characters who work within a Robin Hood–esque political project. No matter what their outcome, Borman understands Cofán struggles according to their inevitable victories: either this-worldly survival and liberation or righteousness in the frame of history and eternity.

Borman is no megalomaniac. Nevertheless, he often depicts Cofán political struggles in dramatic terms, at least in writing and, perhaps, in prayer. The real-time flavor of the conflicts is much more pedestrian. Borman does not constantly proclaim Cofán heroism. Instead, his sense of the novelistic proportions of Cofán politics comes down to the possession of a virtue that is often quiet and subdued: courage. In another passage from the same manuscript, Borman discusses his perspective on this life-defining attribute:

Courage is a strange combination of a matter-of-fact acceptance of a situation mixed with a will to action in spite of the odds. It belongs to no individual or group of people exclusively. It appears in odd places and at odder moments. It is the ultimate starting point for human endeavors, and it is also the last human attribute at the end of many of those same endeavors. It is the stuff of legends, the central core of heroism. But it is also the quiet presence behind the lives and actions of people all over the globe who never get written up, whose matter-of-fact acceptance of overwhelming odds and whose determination to make a go of it anyhow are daily realities.

Borman's evocation of courage and heroics relates to the second literary episode. In this story, Borman articulates two of the central political and existential metaphors of his life: the jaguar and the hunt. Hunting is one of Borman's true loves. He admits that it was his passion for hunting that brought him closer to Cofán people rather than the other way around. For Borman, hunting also means the solitude and freedom of the forest. The natural profusion of the tropical environment stimulates him to imagine the complexity of God and the universe as well as his peaceful immersion in the material world. But his quiet attunement to the forest environment is only part of the joy that he feels as he grabs his shotgun to wander into the forest behind his Zábalo home. As in his rendition of the struggle against the oil industry, Borman narrates hunting stories with the language of danger and battle.

The second literary moment focuses on Borman's description of an episode in which Cofán people trekked into the forest to track a jaguar that had snuck into the middle of the village to kill a pig. There was a general anxiety that an animal with the level of skill and audacity required to do this might decide to kill a human. Cofán people both fear and are fascinated by jaguars. They often interpret their more daring, human-centered exploits as signs of attack from enemy indigenous peoples or tremendously powerful shamans. After he describes the departure of the hunting party, Borman paints a picture of their first sight of the beast:

> And into the sunlight now came the King. What words to describe him in the morning light with the dark dew-covered foliage around him as he came, all gold and ebony, proud with a pride to draw tears from your eyes, strong with a strength that translated to grace and speed and power and all the other things that make this animal almost a god to the people who know it best? Knowing he was the enemy, but in that moment loving my enemy with a love that went beyond his beauty and strength to the spirit that lived in his golden eyes as he approached.

People shoot and miss and chase the jaguar through the forest, as it keeps circling back to retrieve the hundred-pound pig. Alone, the jaguar approaches

Borman, who lets off a shot and hits the animal. It takes off again, with the hunters in pursuit:

> We began to run. People were now closing in from all sides. The jaguar made one last supreme effort—not to get away, even as it was dying, but rather to return to its prey. We were about thirty yards away when we heard two shots in rapid succession, followed by a third. We ran up to find Lorenzo, Atanasio, and Rufino staring at the jag. He had finally stopped, panting and blowing blood, and Rufino and Lorenzo had pumped two slugs through his shoulder. When he had tried once more to get up, Atanasio had shot him in the ear. The King was dead. We all stood, in silent respect, just marveling at the animal we had killed. Soon everyone else who had been in on the hunt was there, and the conversation became animated as we all began to tell our experiences. Laughter, relief, the letdown and the elation that warriors have always felt after successfully defending their homes against an enemy, the emotions and the excitement of such an adversary as this—a happy and victorious hubbub as we looked at the huge paws, the glistening teeth; as we admired the spotted pelt, and studied the bunched muscles of our erstwhile foe.
>
> The return to the village was something out of the ageless and timeless heritage of the human race. Rufino and I carried the jaguar between us on a stout pole. We were surrounded by the rest of the hunters as they cleared the underbrush for us and carried our guns in addition to theirs. As we entered the village, all the rest of the people were there to meet us, and the hunters shouted out our victory. The rest of the people took up the cheer, and the forest rang as we carried the jaguar to the center of the houses and laid it down for all to see.

Borman recounts how people divided the body parts according to Cofán custom, with him receiving one of the eyeteeth, as he had made the first shot. He then describes the jaguar in detail, first covering its physical measurements (230 pounds, seven feet two inches from head to tail, and in perfect physical condition) and then reflecting on his "enemy" in an emotional, spiritual tone:

> Since then I've had to do with a number of jaguars, several bigger than this one, and most of them peacefully. I have enormous respect and love for all of them. Their world is in danger as our human world expands. As a symbol of what we will lose if we lose our natural heritage, they must be preserved at all costs. The other jags I have killed or helped kill have always left with me a deep sense of regret and frustration that it must be so. Strangely, the

exception to that regret is the animal I have just described. This was the King, and we, as humans but as a part of a natural world and rhythm that is as old as time, conquered him fairly and cleanly in combat. And in a strange way, this was what he wanted. To have continued life would have been to know the expansion of the Western world with its roads and plantations and coffee fields and pollution, and all the impersonal, terribly unnatural ways in which civilization destroys the wild. The forest in which we fought is now fields, and a poorly constructed and managed oil well spills its wastes not far from where the jaguar died. In his wisdom, he faced us, his age-old enemy, and fought us, and lost—but ultimately, won. He left his jungle at its prime, while in his prime. We who live on fight our own battle against an enemy that gives no quarter and does not love us, even in our defeat. Meanwhile, he lives on forever in our memories.

In Borman's depiction of the episode, we can identify the mirror sense of his avowedly Christian morality. He is committed to loving his enemy, but with the kind of love perhaps best described by Friedrich Nietzsche. The German philosopher favored the aristocratic virtue of loving one's enemies not for their weakness, but for their strength, which allows for the liberation of will in the tragic drama of battle. After knowing Borman for years, I believe that the desire for strong enemies is one of his most defining characteristics. It provides a basic impulse to his self-imagination and political will.

In Cofán cosmology and moral philosophy, there is a tradition of referring to powerful shamans/leaders as *ttesi* (jaguar) because of their ability to transform into jaguar shape as well as their general strength, courage, and aggression. Although no shaman, Borman also identifies with jaguars. He wears the teeth of a jaguar around his neck, and while painting his face for meetings or religious services, he applies lines of whiskers along his nose. As should be clear in the above passage, Borman both desires and identifies with the fate of the jaguar, whose strength, in Nietzsche's words, "has the look of bronze and knows itself justified to all eternity in its 'work'" (1992:522). After years of knowing Borman, as well as much more time reflecting on his writings, I find it all too tempting to see his choice of taking sides with Cofán people in their struggle against an incredibly destructive world as the path suggested by an overwhelming will to life.

POSSIBILITIES FOR REPRODUCTION

Of course, there are many cognitive factors that underlie Borman's political agency. He is a very intelligent man. His memory is extremely good, and his interest in history is matched by an encyclopedic frame of mind. (Borman joked that I should be careful with my writing, as he remembers all of his statements, word for word.) He also has a gift for reasoning through counterfactuals, which he connects to his love of science fiction and its construction of alternative worlds. Adept at metaphoric extension and imaginative projection, Borman also has an uncanny ability to keep multiple scenarios in mind. As the head of an NGO, he can tell a potential donor what he would do with any amount of money, plotting out a number of courses of action in relation to the amount of resources that would be available for each.

It is hard to draw a line between the cultural and the cognitive. We will never know if Borman's political skills result from his multicultural development, his resultant triculturality, or the inherent quality of his mind. What is clear, however, is that any future Cofán leader who hopes to match Borman's success must possess at least some of his qualities. The most challenging aspect of reproducing his capacities in the next generation of Cofán leaders is the impossibility of re-creating the developmental trajectory of his life. Even if Cofán people do succeed in producing young people with similar skills—whether English speech or scientific training or imaginative project conceptualization or consistent ethical orientation—they cannot hope to make another "gringo chief." Cofán political struggles depend on the aid of important outsiders. For them, Borman is particularly compelling not only because of his supposed transculturation, but because he can appear to be both ego and alter at the same moment.

At the broadest level, the Cofán experiment amounts to a process in which all Cofán people learn to understand their lives in a politicized light, whether through work with NGOs, involvement with conservation projects, or attendance in outside schools. As I explain in later chapters, the deepest challenge to their struggle stems from the polarized nature of Cofán values, which oppose the generalized calm of community life to the egotistical aggressiveness of shamanlike leaders, who are masters at negotiating encompassing realms of violent difference. We can locate the ultimate internal explanation for Borman's efficacy in this distinctly Cofán logic. Unlike some Cofán leaders, he is successfully negotiating the tension between selflessness and ferocity. His moral and cultural balancing act is the essential foundation of his advocacy for the Cofán nation.

2 Identity

One of the most important achievements of the Zábalo Cofán was the ejection of oil companies from their territory in the mid-1990s. After discovering seismic exploration crews on their land, community residents went on the offensive. After a series of encounters, they confiscated the workers' equipment and brought the workers to the village center, where company officials had to retrieve them and negotiate with Borman and other angry Cofán. The most definitive act of resistance, however, occurred after the people of Zábalo found an exploratory well at the southeastern edge of their territory in 1993.

Realizing that their way of life was in danger, residents decided to take over the well in order to shut it down and force a confrontation with Petroecuador, the involved company. With a pair of gringo allies who came along to film the event (and make it available to the international media), a community-wide contingent trudged for days through swampy forest to reach the well. Shortly before arriving at their destination, the Cofán put on face paint, traditional tunics, and palm-reed headbands. They also applied red dye to the tips of their spears. When they entered the site, they moved in and out of the forest to give the impression of a much larger group. They knew that mestizo workers would be terrified to see *indios* emerging silently from the jungle in such a remote area, with spears in hand.

After rounding up the workers, who offered no resistance, the party used the radio to call company officials. Shortly thereafter, a military helicopter arrived with an angry colonel and heavily armed soldiers. At first, the colonel refused to speak to Borman, who was wearing the same clothing and face paint as the others. Then, a Zábalo man stepped up to defend Borman. In broken but angry Spanish, he accosted the colonel. As recorded in the video, which two television documentaries have featured, the exchange occurred in the following terms, which Zábalo residents have confirmed:[1]

> [Narrator:] When the delegation arrived, the chief negotiator was an indignant army colonel with armed soldiers ready for a fight, but not prepared for a gringo chief.

[Colonel, speaking to Borman:] Listen, you are an American. And I'm asking you, Do Americans allow us to come to your country and decide your fate? I will only talk to someone who is a native, with the authority to decide matters that affect the Cofán people.

[Narrator:] What happened next was a defining moment for Randy and the tribe.

[Cofán man:] Hey, listen. He is a native. He was born here in Ecuador. You don't even speak our language. He speaks it fluently. He is a Cofán, and he speaks for all of us.

[Narrator:] Suddenly the army and oil company were facing Cofáns armed with a new weapon: Randy Borman. . . . When Randy spelled out, point for point, the details of Ecuadorian environmental law, it was all but over. You could see it on the colonel's face.

Shortly after the confrontation, the company decided to abandon the wells and offer compensation to the community. As of yet, neither Petroecuador nor other corporations have returned. A large part of the action's success was Cofán people's defense of Borman's identity in the face of doubting outsiders. They had to position him as an authentic Cofán person and a legitimate Cofán leader. Today, outsiders continue to arrive in the community. Most are friendly, but many feel a similar skepticism.

Almost every journalist, tourist, official, and scientist who visited Zábalo during my fieldwork asked me the same question: "Is Randy Borman *really* Cofán?" Some posed the issue in terms of social dynamics, wondering whether the Cofán "accept" Borman as "indigenous" or as "one of them." At the most general level, the answer to the queries is "yes." Cofán people consider Borman to be Cofán. In Zábalo, he fits the pragmatic criteria of Cofán-ness: he speaks A'ingae; he produces his own food through hunting, fishing, gardening, and gathering; and he participates seamlessly in the daily flow of social life. Borman's political position, however, depends not only on what he shares with other Cofán people but on what makes him different. The recognition of his difference is a key element of Cofán politics.

In this chapter, I analyze the logic of Cofán identity in order to shed light on the cultural dynamics of Borman's political position and Cofán mobilization more generally. Borman's case highlights the relationship between power and difference that all Cofán individuals negotiate as they strive to create leaders who can establish a desirable position for the Cofán nation in an increasingly hostile world.

The two terms that Cofán people use to refer to themselves are "Cofán" and *a'i*. The first is a clear artifact of the Spanish Conquest. No one knows for certain what its origin is, but the most plausible theory is that it was a colonial misconstrual of an A'ingae phrase. We can imagine that an expedition descended over the Andean foothills in the sixteenth century and encountered A'ingae speakers on the Aguarico River, one of whose uppermost tributaries is still marked on maps as "Río Cofanes." A'ingae speakers referred to the waterway as Cofa Na'e, meaning "strong" or "important" river. And upon being asked who they were, the proto-Cofán probably replied, "*Cofa Na'e'su a'i*" (People of the Cofa River). Given colonial ignorance of A'ingae phonetics and pre-Cofán identity practices, the explorers rendered the phrase as "Cofanes." Over the next few centuries, the term became the standard label for A'ingae speakers of eastern Ecuador and Colombia.

In the phrase *Cofa Na'e'su a'i*, the word *a'i* probably meant "human being" in the broad sense. The Cofán use *a'i* as a modified noun in a number of identity phrases, grouped in a hierarchy of contrastive sets. A Cofán person living in Zábalo, for example, embodies a whole series of nested identities, all indicated with the term *a'i*:

> *a'i*: human (opposed to nonhuman);
>
> *a'i*: indigenous human (opposed to nonindigenous human);
>
> *a'i*: Cofán indigenous human (opposed to non-Cofán indigenous human);
>
> *Aguarico'su a'i*: Cofán from the Aguarico River (opposed to Cofán from the San Miguel River);
>
> *Ccájeni'su a'i*: downriver Aguarico Cofán (opposed to upriver Aguarico Cofán);
>
> *Zábalo'su a'i*: downriver Aguarico Cofán from the community of Zábalo (opposed to Aguarico Cofán from other communities).

There are also words that derive from *a'i* and refer to liminal states of humanity itself: *aipano* ([unnamed] infant), *Ai'pa* (potentially violent Western Tukanoan person), *ai'vo* (body), *ain* (dog), and *aiñan'cho* (pet, or domesticated animal of any kind). When you ask a Cofán person why his pet squirrel monkey, raised from a baby in his home, will itself eat squirrel monkey meat, he will tell you laughingly, "It has become an *a'i*." In other words, it considers itself to be a person, and squirrel monkeys are no longer its conspecifics.

In some contexts, *a'i* can refer to all indigenous peoples, at least from the lowlands.[2] Cofán people know that Amazonian peoples resided in the region long before colonists. In addition, they realize that other indigenous ways of living are similar to their own. Accordingly, they judge as *a'i* anyone who shares the main elements of their way of life. I asked a young trilingual (A'ingae, Siona, Spanish) man what *indígena* means. In his response, he switched between Spanish and A'ingae and explained, "*Viven de, tsampini. Canse'su, na'e otafani can'jen'sundeccu son indígenas*" (They live in the forest, they are forest dwellers. Those who reside on the shores of the forest rivers are indigenous). When I asked a Zábalo elder what *indígena* means, he said, "In our language, it means *a'i, a'indeccu*.[3] Cofán people as well as Secoya people are *a'i*. They are not mestizos. *A'indeccu*, that's what it says. Napo people, too, are *a'i*. *A'indeccu*—that's our way of saying *indígena*."

Currently, the semantic domain of *a'i* is wider than that of "Cofán." In community-internal discourse, it is very rare to hear the word "Cofán." When making a statement regarding the qualities that differentiate Cofán from ethnic others, a person usually prefaces the utterance with "*A'i* are like this." In most circumstances, people refer to the Cofán ethnic group with *a'i*. Nevertheless, Cofán people have embraced "Cofán" as the term with which they represent themselves to outsiders. In this book, I follow their preference.

THE ETHNIC INTERFACE OF COFÁN TERRITORY

As almost every theorist of ethnicity has argued, the phenomenon operates according to a differential logic. Ethnic identity becomes a factor of social life when individuals and groups from different backgrounds interact with one another and form distinct senses of themselves relative to their others. Contemporary Cofán people conceptualize their identity in relation to a wide assortment of peoples. The array of Cofán people's others composes their version of the "ethnic interface" (Whitten 1981) of northeastern Ecuador.

People from various ethnic backgrounds have inhabited and invaded Cofán territory for centuries. By consent and by force, many of them intermarried and reproduced with people who spoke A'ingae and thought of themselves as Cofán. Today, Cofán genealogies and self-conceptualizations evidence a mixed ethnic heritage. The most frequently encountered nonindigenous others were Cocama (mestizo Spanish speakers from Ecuador, Colombia, and Peru as well as their Spanish predecessors).[4] The group consisted of military forces, Catholic priests, river merchants, hide traders, rubber workers, owners of short-lived haciendas, and, more recently, oil company employees, settlers from Andean and coastal Ecuador, and armed combatants involved in the Colombian civil war.

Mixed in with Cocama were Afro-Colombians and Afro-Ecuadorians, whom Cofán people refer to as Singo or Singo Cocama. Many Cofán laugh about the amorous activities of Singo, who left a legacy of curly hair, dark skin, and broad noses in some Cofán families. After the Singo came the Gringo (Euro-Americans). Gringo arrived as evangelical missionaries, oil company workers, tourists, and scientists. Interactions with people of Euro-American descent increased after the middle of the twentieth century. Shell Oil arrived in the late 1930s, and Borman's SIL-affiliated parents came in the 1950s. Starting in the 1970s, tourists and researchers have been regular visitors to Cofán territory.

At first impression, the arrival of ethnic others appears to have been a true invasion. Cofán people, however, actively pursued ties with many of them, usually for economic reasons. They trekked into the Ecuadorian and Colombian highlands to trade, reaching Quito and Bogotá. They also traveled far downriver to the Amazon in canoes and aboard steamships. A few worked in the bustling Peruvian city of Iquitos, where they married and had families with Cocama or other indigenous people before returning to their home communities, many years later. Contemporary Cofán are proud of their traveling history. For hundreds if not thousands of years, their sphere of activity stretched from the Andean highlands to the main body of the Amazon, perhaps even extending to the point where the great river meets the Atlantic Ocean.

Before colonizers arrived in eastern Ecuador, the Cofán maintained relations with a multitude of other indigenous peoples. They classed the largest body of indigenous others as "Ai'pa," a term whose main referents are now Siona and Secoya peoples. Ai'pa are downriver speakers of Western Tukanoan languages. Cofán people sometimes fought them. At other times, they married them and formed interethnic communities. The term "Ai'pa" has a slightly pejorative sense, meaning "savage" or "killer" in some contexts. Nevertheless, many Cofán people have Siona or Secoya ancestries. When someone criticizes an individual for being angry or violent, they often call attention to the person's Ai'pa heritage. Typically, however, Cofán view Siona and Secoya as equals with a mode of life that is similar to their own. Many Cofán think of them as suitable spouses and coresidents. In previous and current generations, there is a significant degree of bilingualism in A'ingae and Siona-Secoya.

Some Cofán people suggest that Ai'pa are Tetete, a related ethnic term that denotes naked, hostile, murdering savages, the most recently encountered of whom spoke a Western Tukanoan language (Cabodevilla 1997). The Tetete grouping overlaps with warring peoples remembered now mainly through names and downriver locations. One group was the Ikabate, who were probably Western Tukanoans. Another was the Avushiri (sometimes spoken as Avishiri or Awishiri), whom anthropologists have identified as a

Zaparoan-speaking people who occupied the south side of the middle Napo River (Peeke 1973:3-4; Rival 2002:21-33). Through the late 1960s and early 1970s, Cofán people saw unclothed Tukanoan speakers in their territory, usually on the tributaries of the Aguarico. Cofán talk of killing and being killed by Tetete. Nevertheless, there are also strange stories of reciprocity: Tetete sneaking into unoccupied Cofán houses to steal pots and machetes and leaving piles of useful plants in return; or Tetete fishing near Cofán people on the Cuyabeno lakes and preparing mounds of meat for the Cofán to come and collect.

Most Cofán believe that the Tetete have disappeared. Other indigenous peoples, however, now occupy their former territory. The newcomers hail from other areas of Ecuadorian, Colombian, and Peruvian Amazonia. The group includes Shuar, Pastaza Runa, and Witoto people. Nevertheless, its largest segment consists of Napo Runa, or Quichua speakers from the Tena-Archidona area, whom Cofán refer to as Napo, Quichua, or Ingano. Although Cofán frequently intermarried with Napo in the recent past, most now feel extreme hostility toward them. Cofán view them as imperial usurpers who live by drunken, violent, and deceptive ways. Napo communities now occupy much of the Aguarico and San Miguel River valleys. Cofán feel that the pseudo-indigenous invaders have surrounded and dispossessed them.

Further enriching the Cofán ethnic interface is a whole collection of what we might call "supernatural" indigenous peoples. The label, however, is somewhat misleading. Many Cofán believe that these others are just as real, and just as physical, as everyday human beings. They are the "Forest People" and the "Invisible People." Some Cofán shamans claim to have houses and families with them, far away from village centers. Many other supernatural peoples only show themselves at dawn, at dusk, or when Cofán dream or take hallucinogenic drugs.

Although some Cofán elders say that they have seen naked, bow-and-arrow-wielding peoples in the middle of the forest, the majority of hidden peoples appear as Cofán. Nevertheless, they are fantastically adorned and incredibly beautiful. They wear long, clean tunics and intricate feather headdresses and have complexly painted faces and bodies. Some have a moralizing influence on Cofán communities. They urge contemporary Cofán to follow the traditional menstrual prohibitions and live according to Cofán customs on pain of physical or supernatural attack.

In short, Cofán people understand themselves in relation to a wide assortment of ethnic others. Some are friends to be met on the streets of Lago Agrio. Some are bosses or coworkers. Some are enemies to be feared and fought. Some utter questions or insults in languages that are only marginally understood. And others are ancestors, spouses, and kinspeople, living in ostensibly Cofán communities. In all these situations, Cofán people observe, interact with, and form opinions about ethnic difference. In this way, they construct their understanding of themselves as a distinct and definable people.

Explaining the diacritics of ethnic identity, Frederik Barth writes, "The features that are taken into account are not the sum of 'objective' differences, but only those which the actors themselves regard as significant" (1969:14). During my time in Quito, Zábalo, and other Cofán communities, I recorded instances of ethnic identification as they occurred in everyday contexts and interview settings. Together, the episodes evidenced an underlying Cofán philosophy of identity. Although there is variation in individual usage, Cofán people employ a set of shared identity categories to express their opinions about who is or is not Cofán—and why.

The Unchangeable: Body Type and Genealogy

Some features of the Cofán identity field are relatively unchangeable as they occur in concrete individuals. One of the most important characteristics is body type. Relative to others, Cofán people describe themselves as short, dark-skinned, and dark-eyed. Some even posit an identifiable Cofán facial phenotype. They can look at old black-and-white pictures, culled from the records of missionaries and journalists, and pick out the individuals who are and are not Cofán, based mainly on physical characteristics.

Manifesting a certain body type, however, is never a sufficient condition for being Cofán. More often, individuals use it as a negative qualifier to cast doubt on one's identity. For example, when I talked to a young Cofán leader about Randy Borman's ethnic status, the leader acknowledged Borman's cultural and linguistic competence. Nevertheless, he turned to Borman's body as a sign of difference, saying, "But his body is other, he has a Gringo body." Interestingly, one of Borman's own children, himself the product of Euro-American and Cofán-Siona descent, agreed that Cofán are dark-skinned people. He added, however, that his father's skin color—as white as white can be, to my eyes at least—is darker than any Gringo's.

Related to body type as an ethnic differentiator is genealogy: the ethnic types of one's parents and grandparents. As with body type, having a Cofán ancestor is never a sufficient condition for acceptance as Cofán. It can help, however. Individuals who occupy marginal positions in Cofán communities often point to reputed Cofán ancestors as justifications for their inclusion.

For example, a Zábalo family that is typically identified as Napo sometimes offers a genealogical argument for their inclusion. Although they understand A'ingae to varying degrees, they do not speak it. In addition, they do not wear traditional Cofán dress, even on ceremonial occasions. Nevertheless, they often claim that the parents of the group's female head were Colombian Cofán, thus arguing for their rightful place in the community.

When I asked a Zábalo resident about the identity of the family's matriarch, he said that her mother was Cofán and her father was Napo. Consequently, he concluded, "She really isn't Cofán, she's just *ecchoen'cho* (mixed)." As for virtually every other member of the community—all of whom have one or more non-Cofán ancestors—people typically cite their descent from ethnic others in petty squabbles. Almost always, the criticisms are related to conflicts between the questioned and questioning individuals. Normally, if a person follows Cofán norms, neither genealogy nor body type excludes them from Cofán-ness.

The Adoptable: Language, Custom, and Equipment

Unlike genealogy and body type, virtually every other diacritic of Cofán identity can be learned. Some of the qualities serve as signs in a double sense. Not only are they ethnic characteristics in themselves, they also point to the effort an individual expends in developing them. In this way, they express one's commitmeont to establishing and maintaining relationships with other Cofán people—an important indicator of Cofán-ness.

The paramount learnable diacritic of Cofán-ness is speaking A'ingae. The word itself is a combination of *a'i* and an adverbial suffix. "A'ingae" literally means "as *a'i*." In order to learn A'ingae, one must live with Cofán people for extended periods of time. Therefore, speaking A'ingae is both a first- and a second-order sign of Cofán-ness. To use the language is to be Cofán. And to be *able* to use the language indicates a long history of learning based on intimate familiarity as well as the possibility of maintaining that familiarity into the future. Currently, there are no people who speak the language but are not accepted, at least to a certain degree, as ethnic familiars.

A small number of self-identifying Cofán individuals understand but do not speak A'ingae. Their linguistic standing also has a double meaning. Not only does it indicate that they are newcomers or closely tied to ethnic others, it also suggests that they might be ashamed of their Cofán-ness. Their linguistic practice functions as an index of ethnic choice. Cofán people pose the question as, "Why would a person who understands A'ingae not speak it?" And their answer is, "Because they do not want to be Cofán." Refusing to express Cofán-ness is a good reason for ethnic exclusion.

In Zábalo, one young household contains a woman who is unproblematically Cofán, her apparently Napo husband, and their child, who understands A'ingae, Spanish, and possibly Quichua. The father comprehends A'ingae but does not speak it. It is common to hear the couple chattering back and forth, wife speaking A'ingae and husband speaking Spanish. One Zábalo resident reflected on the husband's identity:

> He lies and says that he's *a'i*. "I'm Cofán," that's how he lies. And so another *a'i* from a different place comes and asks him, "What kind of person are you?" And he says, "I'm Cofán." So another *a'i* then says, "Hey, here's a speaker, speak to him." And you go and speak to him, and he can't speak. Well then, you're a liar. You're really a Napo! You're not our people, not Cofán.

Reflecting on the same individual, another Zábalo resident said, "When he speaks our language, I, too, am going to think that he's already *a'i*, Cofán, when he speaks in A'ingae. For him, only the language is missing. Only that is missing for him to be more Cofán."

In addition to the principal ethnic sign of language, Cofán people identify a whole series of practices as *a'i canse'cho* (Cofán customs). Some of the most commonly cited *a'i canse'cho* include: traditional dress (home-sewn tunics for men and blouses and skirts for women); traditional food (*cui'ccu* [banana drink], which is the main form of Cofán carbohydrate consumption, as well as stews made of game meat and boiled green plantains); certain forms of hunting and gardening; music and dance styles; traditions of shamanic activity and knowledge/power; marriage practices; and ritual activities and festivities. Even objects become markers of Cofán identity. A popular genre of humor entails referring to a typically Cofán item as an object from national Ecuadorian culture: an axe is an "*a'i* chain saw"; an old stick used for measuring lengths of palm wood is an "*a'i* meter"; grated plantain is "*a'i* rice"; and the ever-present banana drink is "*a'i* coffee."

The Implicitly Shared: Forest Life and Peaceful Sociality

Connected to the assortment of *a'i canse'cho* is a more encompassing but less explicit sense of being people who live in and with the *tsampi* (forest). To a large degree, being Cofán means being a *tsampini can'jen'su* (forest dweller). Cofán people use their dependence on the forest to differentiate themselves from such peoples as Cocama and Napo. Zábalo residents assert that many ethnic others embrace a commercial and predatory relation to the Amazonian environment, which they believe to be very different from their own.

Being a *tsampini can'jen'su* is not a spiritual or cosmological position. Rather, Cofán people explain it as a preference for producing one's means of life through self-sufficient, non-market-mediated activities, which depend on residence in a relatively intact environment. The Cofán describe an intensively market-mediated lifestyle as "buying to live." They point to how Cocama and Napo want to live along roads so that they can transport their crops and be within reach of urban centers. My Zábalo consultants said that they did not want to live along roads for a number of reasons: absence of game, antagonistic relations with landowners who will not allow Cofán to hunt on their holdings, and outsider desires to steal Cofán canoes and dogs.

In addition, Cofán people identify themselves by the joy and knowledge that are part of their forest-based lifestyle. The following passage is one of my favorite articulations of this sense. The speaker is Borman's oldest son, who at the time was an excitable trilingual (English, Spanish, A'ingae) fourteen-year-old. I asked him what a Cofán person would be if they did not enjoy hunting. In English, he replied:

> You can't be grossed out by dissecting all the stuff you kill, and killing and blood dripping all over you. If someone can't do this, then they're lazy. They're a Cocama or a Spanish person or an English person, a Gringo. Because they're lazy. And they're grossed out so they won't do that stuff. They'll be at desks working and not really going out and having fun killing stuff.

Cofán girl with her pet monkey clinging to her hair (Photo by Michael Cepek)

Connected to a forest-based lifestyle is a wider sense of positive sociality. Ideally, Cofán communities are composed of peaceful, reciprocal, and symmetrical relationships between individuals and family groups. Propensity to violence, anger, or disruptive behavior of any kind is cause for declaring a person *fuesu a'i* (other people). Cofán oppose their preference for a calm, peaceful, and generous lifestyle to the aggression, hostility, and stinginess that they attribute to every other ethnic group with whom they maintain relations.

When Cofán people describe the calm conviviality that prevails in a "good" community, they use the word *opatssi*.[5] The term refers to a socioexistential state of freedom from worry, fear, and need. More positively, it connotes a condition of health, strength, satiation, and peace. Anyone who is incapable of contributing to a generalized *opatssi* state risks ethnic exclusion. When a longtime Zábalo resident got drunk and fought with his family, for example, other Cofán pointed to his father's Siona descent and declared him Ai'pa. A few older Cofán remarked that as soon as the man and his brothers begin to drink, they pick up their palm-wood spears and get ready to kill, as all Ai'pa are prone to do. In normal circumstances, however, no one questions the man's Cofán-ness.

Living in an *opatssi* state with other Cofán individuals means living with relatives. If people in Cofán communities are not already united by family ties, they marry their children to each other or establish ritual kin relations. Virtually every Zábalo resident has kin terms with which they address and refer to one another. As relatives, individuals are bound by expectations of cooperation and material reciprocity. If someone exhibits greed or selfishness, others can ostracize and even "de-Cofanize" them.

When a Zábalo man found employment with an oil company and did not sufficiently share his earnings or provide work opportunities for others, people referred to him as a Gringo, even though he has no cultural or biological ties with Gringo people. In the same way, people who give, share, and help others often receive the title "good *a'i*" or "old-time *a'i*" (i.e., *really* Cofán people). When talking about a Zábalo man with strong Napo ties who tries to keep others from hunting near his home, a resident said, "That man is an unthinking fool. He doesn't think Cofán thoughts. He's a Napo, and those are Napo thoughts and desires."

Negative Identity

There is an additional sector of the Cofán identity field whose content is more difficult to uncover in an interview context. In some circumstances, Cofán people identify themselves by a set of "negative" characteristics. Many of their embarrassing incapacities are objects of humorous, self-deprecating commentary. Examples include: not being able to stand the flavor of garlic, cheese, or beef; not remembering to say *perdón* after acts of flatulence or eructation in the presence of non-Cofán; not being able to walk in Quito without getting lost; not knowing how to cook vegetables and other bought food items; and not being familiar with the basic objects of urban environments.

For example, during a visit from a remote lowland community, a group of Cofán stayed at the Quito Cofán Center and decided to wash their clothes in the toilet bowl. The event caused hours of good-natured ribbing by the more cosmopolitan resident Cofán, who joked with the visitors for being "truly Cofán." Humorous, self-deprecating discourse is a marked form of Cofán sociality. It evidences awareness of a certain cultural and historical moment in national and global contexts as well as a lasting confidence and satisfaction with a forest-based way of life.

Other negative qualities of being Cofán are causes for real anxiety. Examples include not being able to speak Spanish, count, or earn money. Some of the more serious negative aspects arise in speculations concerning atypical individuals, such as Randy Borman. Cofán people ask themselves and each other, "How could Borman be Cofán, given that he possesses and reads hundreds of books, that he knows how to write grants and run an NGO, and that he can find enough money to live in Quito, even if he does not want to?" Clearly, Cofán people would like to be able to do what Borman does, even if they do not want his non-Cofán capacities to replace the skills that enable their forest-based existence. Many of the current educational projects of the Cofán attempt to create individuals who can replicate at least some of Borman's abilities. In this sense, the structure of Cofán identity is not static or constraining. Rather, there is an active effort to add essential elements to it without losing what it already indexes.

In summary, the Cofán philosophy of identity is composed of multiple, overlapping categories: the set qualities of body type and genealogy; the adoptable aspects of language, custom, and equipment; the more implicit conditions of forest life and peaceful sociality; and the negative aspects of being Cofán, some laughed about and lived with, others the objects of intentionally transformative projects.

CHANGING IDENTITIES

Cofán identity is not a simple, static, all-or-nothing game. People disagree about who is or is not Cofán. At different points in time, and in different social situations, a person may or may not identify herself or another individual as Cofán. One young man with a (normally) Siona-identifying father and a (normally) Cofán-identifying mother told me that he is Siona when he is in a Siona community and Cofán when he is in a Cofán community. Reflecting on the conundrum, he suggested that he is "both." With regard to Borman's ambiguity, the same individual usually calls him an *a'i*, although I have also heard him describe Borman as an "*a'i* Gringo." To put an end to the recurring conversation, the young man asked me what Borman calls himself. I replied, "*A'i.*" He then said that the matter is settled: he is clearly Cofán.

People claim that some individuals are more Cofán than others. Given variations in phenotype, genealogy, language ability, and lifestyle, an individual's Cofán-ness can be qualified, or, to be more exact, quantified. During my fieldwork, I heard a number of phrases that indicate different levels of Cofán-ness: "Cofán Cofán" ("a Cofán's Cofán," to use an English idiomatic gloss), "*ña'me* Cofán" (truly Cofán), "*tittse* Cofán" (more Cofán), "*verdaderamente propio* Cofán" (truly properly Cofán, in Spanish), "*completamente* Cofán" (completely Cofán, in Spanish), "*medio* Cofán" (half Cofán, in Spanish), "*más* Cofán" (more Cofán, in Spanish), "a little bit Cofán," "really Cofán," and even "I'm more Cofán than you [Mike Cepek]." (One of Borman's sons uttered the last phrase when I protested that I should be on the Cofán team in a volleyball game rather than him, whom I had placed with the gringos). Borman's parents, who speak A'ingae and practice many forms of *a'i canse'cho*, told me, "We're quite Cofán. We've got quite a bit of it [i.e., Cofán culture] in us." Cofán people concur. When a small child and her mother happened upon Bobbie Borman in a Cofán community, the child, who did not know her, pointed and said, "Cocama." The mother corrected her, saying, "No, *a'i.*"

The complexities of shift, mix, and level are related to a key feature of Cofán identity: the idea that both their own and other ethnic allegiances are substantive yet open. In other words, Cofán people understand that one's ethnic identity points to the most basic aspects of their life, such as psychology, language, custom, and sociality. Nevertheless, they believe that a person can transform their mode of being, often through intentional, labor-intensive processes of socialization.

Like the pet squirrel monkey, human beings from different ethnic backgrounds can "become *a'i.*" In this context, the phrase refers not to becoming human, but to becoming Cofán. Inversely, originally Cofán people can transform into ethnic others. One Cofán man who worked for an oil company in his youth told me that he "became a Cocama." He "lived according to Cocama customs" and was guided in his actions by "Cocama thoughts and desires." Nevertheless, after his oil work ended, he returned to a Cofán community and married a Cofán woman. In short order, he "became Cofán again."

Many Cofán people are happy to let other kinds of people live in their communities, but only with the understanding that they will become *a'i* by doing so. People in Zábalo lament the fate of the community of Dovuno, which Napo people have taken over. Old Cofán swear that when the first wave of Napo immigrants came to the Aguarico River in the latter half of the twentieth century, they promised to marry Cofán people and assimilate to Cofán ways. At first, many of them did learn A'ingae and live by Cofán customs. Nevertheless, as more and more of their relatives arrived from the Tena-Archidona area, they changed their behavior. The newcomers stopped learning A'ingae, and the village gradually came to assume the shape of a Napo community. Today, most Cofán suggest that over half of Dovuno's residents are Napo rather than Cofán. Reflecting on the experience of Dovuno and his own community, a Zábalo resident explained a contemporary Napo family's permission to move into Zábalo in the following terms: "Like this they spoke. They will enter the community, and live here, and they will become *a'i*. They will be *a'i*."

With a different use of the same logic, Zábalo residents accepted and then rejected the Cofán-ness of another man. His father is Siona, fluent in A'ingae, and married to a Cofán woman. He has lived in Zábalo since late childhood, and he has a house full of A'ingae-speaking and Cofán-identifying children. Nevertheless, villagers openly questioned his identity after he committed a violent act. Rather than relegating him to Ai'pa status because of his partial Siona descent and propensity to violence, however, they concluded that he was an *a'ive dambi'choa* (literally, one who has not become Cofán).

Finally, to use an example that is much closer to home, one could consider the ways in which Cofán people interpret my own identity. In a common stance toward Gringo researchers, Cofán people accepted my proposal to live and work with them on the condition that I would provide substantive reciprocity. Unlike other Gringos, however, I continue to return to Cofán communities as often as I can. I show a sincere interest in maintaining social relations and participating in the Cofán way of life, to the best of my abilities. Eventually, Cofán people began to refer to me as *a'i*, often jokingly, but sometimes with more seriousness.

I can say with relative confidence that there is not a single Cofán who would call me "truly Cofán." Nevertheless, a significant number of people call me *a'i*. The most frequent instances in which people say, "Mike, you are *a'i*" or "Mike, you are becoming *a'i*," involve their judgments of my increasing proficiency in A'ingae. Nearly as often, Cofán call me *a'i* when

I do things that Gringos simply do not do. Examples include: stirring and consuming banana drink without aid or supervision, eating food without silverware or a plate, showing true excitement at the prospect of a meal of excessively fatty woolly monkey, swinging a machete at a collective work session, going alone into the forest to carry cut wood back to the village, and even making vulgar jokes in A'ingae about male and female anatomy at drinking parties.

Some of the instances depended on my performance of prototypical *a'i canse'cho*. The majority of them, however, involved implicitly ethnicized practices that only appeared *a'i* against the expectations of what a true Gringo would have done in similar situations. And all of them, I wager, were interpreted as good-humored attempts to live as a Cofán. Cofán people recognize my efforts to enter into the web of sociality that defines Cofán life. They interpret me as someone who wants and likes to be with Cofán people and thus might be a suitable addition to their communities.

I believe that my apparent happiness in Zábalo lay behind the offers I received to marry young Cofán women, usually at the insistence of their brothers, sisters, mothers, and fathers. Of course, my resources and knowledge of outside ways go a long way toward explaining the interest in making me a relative and keeping me around. But more significantly, I believe, Cofán people made the offers because they thought that I might reply in the affirmative. They assumed that I might not be so different from themselves, given all the aspects of their way of life that I have managed to learn and enjoy and perform, albeit clumsily. Although I never married a Cofán woman, I did become a ritual relative to families in Doreno, Dovuno, and Zábalo.

The willingness to ethnically incorporate others takes us to the heart of Cofán identity. At its most basic and socially consequential level, Cofán individuals judge as *a'i* anyone whom they can imagine residing in their communities in a satisfactory manner. Cofán communities are not large. Frequent face-to-face interaction is not only the collectivity-bonding stuff of laughter and mutual aid. It is also the similarity-demanding ground for a co-constructed lifeworld, which is extremely sensitive to disruptive individuals whose psychological states, linguistic capacities, practical customs, and social norms can so easily be judged "other."

Cofán understand that one can only be *a'i* by becoming *a'i*. Cofán-ness is the product of a gradual developmental process that one can accomplish only in a community context. One's ability to develop the identity depends on collective permission to live with Cofán people in a Cofán village, which means that Cofán people must recognize the individual's sincere desire to do so. In recent years, the idea of collective permission has become more consequential. In order to acquire a land title, a community must construct a formalized governing structure as part of its legal personhood. A necessary element of the institutional form is the maintenance of a full list of "legal" community members. Although communities have some leeway in determining who can join, current residents must approve any newcomers through a binding vote. If an ambiguous individual wants to join a Cofán community but appears problematic, the *comuneros* can simply refuse his or her petition to become a member. In other words, they can prevent the individual from becoming *a'i*.

ETHNICALLY AMBIGUOUS LEADER FIGURES

Over the last century, Cofán communities have been fragile, short-lived, and amorphous things. With the colonization of northeastern Ecuador, however, Cofán people had to aggregate themselves in more durable, state-sanctioned formations in order to obtain land titles. Nevertheless, Cofán households and residential groupings are still relatively autonomous. Overarching political structures continue to be weak. Many family groups understand collective unions in terms of the immediate benefits of coresidence and cooperation. Sometimes, however, collective living is more of a burden than a blessing. Traditionally, separate households congregated around *na'su* (chiefs) in order to profit from their ability to negotiate beneficial relations with external domains, whether of the ethnic, political, or supernatural kind.

Historically, chiefs could *canqquiañe* (make a community) because they possessed forms of power and knowledge that differentiated them from their followers. Their abundant abilities provided a positive reason to form a settlement. If a potential *na'su* did little to contribute to collective welfare, there was no reason to gather around him. Hence, Cofán people's commitment to organized life and authoritative power is not based on a Hobbesian vision of human nature. Family groups do not constantly war if a Leviathan-like force is not present to discipline them. Rather, Cofán people find the most pressing reason for collective living in protection from external threats, such as raids from other peoples and consistent shamanic attack. A large part of a *na'su*'s obligation is to negotiate the dangers and promises of encompassing domains of difference that trouble community life from the outside.

Currently, communities have an institutional permanence that they did not have two generations ago. Yet *na'su* continue to provide the basis for a community's functioning as a social collectivity. Their cohering power depends on their difference from their followers. Their specificity, which is evident in their ability to interact successfully with external others, makes virtually all *na'su* ethnically ambiguous. Leaders exhibit psychological, linguistic, practical, and social specificities that set them apart from the majority of community members. Cofán people expect their *na'su* to be aggressive; to enter fearlessly into conflict-ridden situations; to lead a life full of worrisome preoccupation rather than *opatssi* well-being; to speak languages other than A'ingae; to possess sufficient resources to travel to faraway places in order to interact with ethnic others; and, in general, to know the norms governing extra-community life, which they must master in order to secure the welfare of their communities.

Although Randy Borman is perhaps the most extreme example of an ethnically ambiguous Cofán leader, other *na'su* rival his peculiarity. One of them, a longtime president of both FEINCE and his home community, has no Cofán ancestors. By descent, he is completely Cocama. He understands bits of A'ingae-language discourse, but he does not speak the language. His wife is Cofán, and his children are bilingual and identify as Cofán, at least in some contexts. This leader stresses that he has lived in a Cofán community for most of his life. He asserts that his social and residential history, as well as his appreciation of community life, do not make him an "other" from the Cofán perspective.

Zábalo's current president is not ashamed to say that his father was a Cocama. His Cofán mother conceived him with an oil worker, whom she did not marry. Subsequently, she and her Cofán husband raised him as their own. The man maintains relations with his Cocama kin, who live in the distant town of Coca. They often aid him with political, economic, and educational support. He looks noticeably different from most Zábalo residents, given his tall height and light skin tone. Moreover, his year in the Ecuadorian military and his relations with non-Cofán family members make him the most adept Spanish speaker in Zábalo, aside from Borman. Despite his ties to ethnic others, he is the most vociferously proud *a'i* in the community. He speaks A'ingae with a speed and dexterous wit that few can match. Even more importantly, he loudly proclaims his love for life in Zábalo and his willingness to fight to the death for his community and its territory.

One important young leader, who holds positions in FEINCE, the FSC, and the community of Doreno, told me about his desire to learn the ways of ethnic others. Not only does he want to master Spanish so that he can work with Cocama and make use of the national media, he also wants to learn Quichua, Siona-Secoya, and Huaorani. He has an inherent interest in other peoples and languages. In addition, he believes that intercultural skills will help him in his work. Nevertheless, he realizes that a disciplined commitment to learning the languages of other peoples, especially through formalized education, will complicate his Cofán-ness.

This leader related his desire for learning the ways of others to the objective of Cofán education projects, which aim to produce Cofán leaders with literacy skills in Spanish, English, and other languages. He suggested that the students will lose part of their Cofán-ness in the process, hence risking transformation into ethnic others of unstable allegiances. In describing the possibility, he referred to a former president of FEINCE. The past leader has a relatively pure Cofán genealogy, and his proficiency in A'ingae is beyond question. Nevertheless, he no longer participates in many of the *a'i canse'cho* that compose Cofán-ness:

> Yes, the Cofán children in outside schools will become like that. That's what I think. If they study. [But is that a good thing?—M. Cepek] It's good. As political leaders, when they become *na'su*, I want it to be that way so that they have a lot of experience with other lands and places. [But will they lose their Cofán customs?—M. Cepek] No, look at it like this. The FEINCE president can't live by *a'i* customs. He doesn't live an *a'i* lifestyle. He reads books and papers every day. And he writes. And he lives with Cocama people in Lago Agrio. He doesn't have a garden, and he doesn't go hunting. He can fish, but he doesn't know how to shoot a gun. That's what I'm talking about. That's what these kids will become. But that's all right, if they do it to help *a'i*.

The young leader concluded by saying that sociocultural transformation is acceptable as long as Cofán youth work to change themselves in order to become *na'su* and help the Cofán nation. People recognize that community life is made possible by the work of exceptional individuals who have the power to transcend political, ethnic, and supernatural boundaries. *Na'su* stretch Cofán-ness to its furthest point. People claim them as Cofán not only because they participate in the Cofán way of life—a participation that may be atypical, and even problematic—but because they actively work to make Cofán existence possible in an increasingly threatening world.

At this point, we can return to Borman's reflections on the first time Zábalo residents "forced" him to wear Cofán regalia in a public setting. The incident occurred during the campaign against the oil industry in the early 1990s. In media events and official meetings, Cofán people wanted outsiders to interpret Borman as their legitimate spokesperson, and they made sure that he wore his tunic, headdress, and face paint. Their decision reflected their long-held understanding of Borman as a Cofán person. Nevertheless, the very act of donning his ornamentation and taking on the role of a visible defender of Cofán rights increased Borman's Cofán-ness. The question of whether Borman's identity play was "strategic" or "authentic" is entirely beside the point. A central element of Cofán-ness is the desire and ability to sustain a satisfying quality of life in Cofán communities. Cofán people's decision to claim Borman's identity during an important political event simultaneously presupposed and produced his Cofán-ness.

BORMAN FROM THE COFÁN PERSPECTIVE

Although most Cofán people accept Borman's Cofán-ness, they also stress his uniqueness and ambiguity. Rather than summarizing Cofán evaluations of Borman's ethnocultural status, in this section I provide a set of direct statements on the issue. My translations of the A'ingae are a bit jerky, but I want to communicate the logical flow of the original assertions.

When I ask Zábalo residents whether or not Borman is *a'i*, they typically describe his ability and desire to participate in the Cofán way of life. In the first passage, a Zábalo elder talks about Borman's childhood in Doreno, including how other Cofán people defended his identity against critiques from doubtful outsiders:

> When in Doreno, Randy was just a small child. He grew up and he spoke all of our language. He understood well. And he would say, "I'm *a'i*." That's what he would say. The Cocama didn't like that. They would say that he's a Gringo, that he's not really *a'i*. That's what the Cocama would say. That he's a Gringo, that he's a liar. They used to speak like that. But then we said, "No, he's not a Gringo. He's *a'i*. We raised him from when he was a little child, really small like this [raising his hand a couple feet off the floor]. That's why he's helping us." We'd say that, and the Cocama would stop and think and not know what to say [laughs]!

The next two passages, recorded from different informants, explain Borman's mastery of a forest-based lifestyle, an essential condition of Cofán-ness:

> A'i people live without buying anything to eat. They just make fields and grow manioc, plantain, and banana. They harvest those and make their drinks. Then they drink and live. Randy knows how to do this. Like that he grew up with us in the forest. He grew up hunting, and he didn't buy anything. He would bring back game and eat. He knows how to do that.
>
> Randy is like this. From his childhood he knew that he grew up here, and he knew how to eat forest meat. And he knows the forest. He knows everything, the way that everything is. He knows more than most *a'i*! He knows how to make canoe paddles. The young people of today don't know how to do this. Randy is more knowledgeable about this and many other things.

In the next statement, a Zábalo resident refers to Borman's *a'i in'jan'cho* (Cofán way of thinking and desiring) as part of his identity. The main point is his desire to live in an intact Amazonian environment:

> His thoughts and desires, those are *a'i*. That's why he hates the idea of destroying the forest. He grew up with *a'i*, and he has *a'i* thoughts and desires. He doesn't want to destroy anything. So that forest medicines don't disappear, so that we don't dirty our streams and rivers, so that we can eat and drink and nothing bad happens to us. That's the *a'i* way of thinking and desiring.

Some Zábalo residents even describe Borman as an "old-time" or "traditional" Cofán. One man referred to his ability to conduct marriage ceremonies according to Cofán custom. The individual even compared him to Guillermo Quenamá (whose nickname was "Yori," now modified by the -*ye* death/honorific suffix), Doreno's founding *na'su*:

> Randy knows the old-time ways. That's the way he thinks. He has all of that in his head. Everything from the old times. He even knew Yori'ye. He knows. He knows how to marry people. How to scold and council people. When Randy is marrying people, he will really advise them. He's like the old-time people. That's what he's like.

Using the same language of custom, Cofán people sometimes distance Borman when they describe his ability to maintain multiple lifestyles. They realize that he can move between identities in a way that most Cofán cannot. In some circumstances, people see him as a fluid, shifting, and mixed individual. Nevertheless, his movements do not preclude his Cofán-ness:

> If he's here in Zábalo, he lives like an *a'i*. He drinks banana drink, he eats forest meat, he drinks manioc beer. He works his field, he makes his own house, he goes hunting. That's how he lives. When he goes to Quito, he lives by other customs. That's why we say he has Gringo thoughts and desires. Those really aren't *a'i* customs.
>
> When he comes here he's like an *a'i*. But when he goes to the land of the Cocama he lives according to Gringo customs. So it's like he's an *a'i* when he comes here. Yes, he lives like an *a'i*. When he comes to the forest.

Another Zábalo man expressed a similar sentiment. In his telling, however, Borman's original identity is Cofán, and he is now in the process of becoming Gringo: "Randy is *a'i*. He's knowledgeable about *a'i* customs. He knows the *a'i* life. But yes, he's Gringo—but now. He went away [i.e., to do political work in Quito] and now he's in the process of becoming a Gringo. Before he resided here."

When individuals reflect on Borman's lifestyle shifting, they sometimes suggest that a true Cofán cannot engage in the same kind of constant and capable fluidity. In this sense, Borman makes people think about the negative aspects of being Cofán: the characteristics that take the shape of incapacities, and that people hope to change, at least in their children. While reflecting on their own challenges, Cofán people tend to question Borman's identity. One Zábalo resident commented on Borman's urban and political skills, wondering how he acquired them, and what they say about his identity:

> Is Randy really *a'i*? Has he really become *a'i*? He knows *a'i* customs. He lives like an *a'i*. Yes, he can live an *a'i* lifestyle. If he comes to Zábalo, he can live here forever. But he also lives in Quito. Because of that, I think that he really isn't *a'i*. If you're like an *a'i*, you wouldn't be able to live in Quito. Let's say that I'm like Randy, and that I leave here to live in Quito. So I go to Randy's house in Quito, and how do I know where to look for money? How to buy things to eat? How to help the kids going to school there? I can't—that's what it means to be *a'i*. So Randy, well, he grew up in school. He has school-knowledge thoughts and desires, and *a'i* thoughts and desires. He has two modes of thinking and desiring, so he knows more. He's more of a thinking person. Like that he lives. He knows everything from school, and everything from here. That's why he knows more.

Reflecting on the source of Borman's atypical knowledge and ability, another Zábalo man wondered why he can do what others cannot, and how his abundance of knowledge makes him something other than an *a'i*:

> Randy thinks and knows how to do everything. How to secure land to live on. He has that kind of experience. *A'i* really aren't like that. We've thought, "What, is it because he has studied in school that he thinks like that?" He can go to any Cocama meeting and speak correctly about the really important things. We can't think like that. So I thought, "Is he just truly knowledgeable, does he think more?" It's a necessity to know how to speak like that. So if we ever go to a meeting, and we're lacking our hearts [i.e., are not thinking well], we know to ask Randy, "Randy, how must this be done?" "No, like this, like this it must be done." He'll explain things more clearly. So we become really happy. "Oh, that's how our hearts were lacking," that's what we say.

Another Cofán man reflected on the same question: What accounts for Borman's knowledge, and how do his abilities differentiate him from other Cofán people?

I think about this with another man from Zábalo. Randy's really like you people. Every day he's reading. Every day he looks at books. Every hour and every minute. Living like that, he knows the customs of all people. Every land's style of life. So now, he knows *a'i* customs, and he also knows Gringo customs. A lot of cultures ["*culturas*" in Spanish]! With all of that, he thinks big. How to do everything. But us, well, I talked about that to Randy's mother. She asked me about the Bible, and I said that *a'i* customs are really difficult. It's really hard to read the Bible every day. We really can't. *A'i* customs are like this—well, you've already seen *a'i* customs, living here in Zábalo. A person here gets up early and drinks banana drink. Then they go to cut bananas, or clear their fields, or fish, or go hunting, every day. Once in a while they might just stay in their house, to sit around and do a little work on their house. That's what *a'i* know. And yes, there's time, but we don't have the interest to read books. We really don't like to sit in our houses every day. That's not *a'i* custom. *A'i* custom is to work. [But don't you want to be a little more like Randy in this regard?—M. Cepek] Yes, I get sad about it, but I don't have that. I don't know how to do that, to just sit like that. I get lazy! All *a'i* are like that. That's why I want some children to become like that. To have that kind of vision ["*visión*" in Spanish], to read and look at books.

In addition, Borman's perceived wealth sets him apart from other Cofán. Although he is clearly not wealthy by North American standards, Cofán people point out that Borman has more commodities than they do. Cofán people's sense of their relative poverty is also an aspect of their negative identity. One man explained, "Our thought is like this. Randy is *a'i*. But we also think that he's Gringo. 'He has an outboard motor, he has everything,' that's what *a'i* think. He really isn't poor. It's like he's not really *a'i*. People who aren't really *a'i* have things—they're Gringo people. That's what we think. *A'i* are poor [laughs]!"

A relatively young leader who is earning a sizable income with the FSC echoed the sentiment about Borman's relative wealth. Nevertheless, he followed it by affirming Borman's *a'i* way of thinking, and especially his desire to help Cofán people. In the end, the man declared Borman a hybrid creature:

A true *a'i*? He's not really like a true *a'i*. I'd say that he's not a true *a'i*, not really like us. His customs are a little different. A little Gringo, a little *a'i*.

Look at it like this. Go to his house in Zábalo. He has a stove, and Cocama food, and everything. *A'i* have nothing. Truly, though, he's *a'i*. His way of thinking and desiring is all *a'i*. And his way of desiring to help *a'i* people. That's really great. But his customs haven't truly changed. A little bit of his customs remain the same, his Gringo customs. They exist together with his *a'i* customs.

I collected the statements in this section from individuals of different ages, social positions, and ethnic complexities. All of them, however, normally identify as Cofán. They express agreement on the basic points. Borman has the ability and desire to live a forest- and community-based life, and he actually lives this life (although now only periodically), thereby making him an *a'i*. Nevertheless, his knowledge of other people's speech and culture, his ability to negotiate external relations, and his possession of the resources to live as an ethnic other complicate his identity. In the words of Cofán people themselves, Borman is not simply different from an *a'i*—he is something *more* than an *a'i*. His superabundance of knowledge and resources makes him an ideal leader capable of sustaining community life in a very uncertain world. As one young Zábalo resident said:

> Randy is a real helper. We alone can't handle the community, where to go, how things must be done—we can't do that. [What do you mean?— M. Cepek] It's like driving/directing,[6] do you understand? A driver/director. We can't do it alone. Randy really helps. He knows. If he asks us about how to do something, we lack an answer, how one has to speak about something. We don't know. Randy thinks more about how we must go further. He thinks that and he knows how to speak. I don't know Randy's thoughts and desires. Randy has a really big head [laughs]! He knows, and he thinks and desires well.

RECONSIDERING COFÁN IDENTITY

Cofán identity involves many of the features of history, kinship, culture, homeland, and ascriptive labeling that anthropologists consider to be central to ethnicity. Although the practical complexities and internal fragmentations of the Cofán *ethnie* might seem legion, the Cofán nation does exist. Being *a'i* is a practical mode of personhood that is substantive yet open. Individuals can enter and exit Cofán-ness in a way that transcends the distinction between constructivist and essentialist approaches to identity. The possibility of transidentification depends on an individual's desire and ability, as well as a community's collective permission, to learn how to perform Cofán life in a Cofán setting. In other words, Cofán identity is constructed, but only through a long-term process that transforms one's essential being.

At the margin and the center of the Cofán identity field are ethnically ambiguous leaders. They provide the enabling conditions for community life by participating in that which exists outside of, and often in opposition to, an ideally seamless internal conviviality. Their position is complex. On the one hand, ethnic openness and external threats prime Cofán people to embrace the political resources of ambiguous *na'su*. On the other hand, atypical leaders can only sustain community life because they possess the skills, dispositions, and resources that make them similar to ethnic others, and thus potentially problematic community members.

Borman's perceived wealth and extraordinary abilities do stimulate jealousy and resentment in some Cofán people. Nevertheless, his political contributions, his appreciation of Cofán culture, and his performance of peaceful sociality make him an acceptable community member, and thus a real *a'i*. To end a long story with a short list of contradictory points: Yes, Borman is Cofán. He is Cofán because of what he shares with Cofán people, and because of what makes him different. He is Cofán because he stands both inside and outside Cofán communities. And he is Cofán not simply because of his similarities to other Cofán people, but because he has become the prototype of the ideal Cofán *na'su* in the age of global politics. By reflecting on his persona and power, Cofán people debate who they are and who they might become as they struggle to overcome their challenges.

3   *Value*

THE DILEMMA OF BEING COFÁN

R andy Borman's uniqueness is only one aspect of contemporary Cofán diversity. In Zábalo, the lives of two men—Antonio and Pablo[1]—express a more basic distinction. Now in his fifties, Antonio is hardworking and generous. I jokingly call him *yaya* (father), as I always eat at his house while visiting the community. During my early months in Zábalo, I gravitated toward Antonio and his wife, Saura, because they were such a pleasure to be around. Antonio frequently brought back a peccary or paca from the forest. Saura prepared delicious stews and giant pots of sweet, warm banana drink. After mornings or afternoons of intense labor, we sat next to their hearth and traded jokes, hunting stories, and reflections on our dreams. Antonio's home is filled with laughter, plentiful food, a pleasing balance between work and leisure, and an overall sense of friendly sociality. I miss it every time I leave.

Pablo is a few years older than Antonio. He, too, offers a bowl of manioc beer or a plate of meat when I stop by. Like Antonio, Pablo is a bawdy joker. We enjoy trying to outdo each other's crassness with exaggerated tales of sex and bodily functions. Unlike Antonio, however, Pablo has a reputation for latent hostility. Zábalo residents say that he is the only person in the community who possesses shamanic weaponry. While I was interviewing Pablo about shamanism, he imitated the process of transforming into a jaguar. His fingers became fangs, and a deep, unsettling growl emerged from the depths of his throat. Pablo laughed deviously when talking about the Cofán of yesteryear's disciplined, violent ways. He recounted their massacres of enemies with spears, hardwood swords, slings, blowguns, and ingenious strategies involving ambushes and trenches. His eyes lit up when he exclaimed, "Old-time Cofán were really, really bad." Pablo is proud of Cofán aggressiveness, even if it exists now in the subdued realm of shamanism rather than the overt violence of warfare.

Although Antonio and Pablo are friends and coresidents, their divergent self-understandings reflect an important distinction in Cofán society. Antonio's sense of the good life is representative of the majority of Cofán people, who prize humor, reciprocity, calm, ease, and good-natured hedonism. Pablo shares many of Antonio's satisfactions, but he believes that ferocity, conflict, and discipline are central to an admirable existence. The distinction between Antonio's and Pablo's understandings of a worthwhile life overlaps with a series of important oppositions: present vs. past, follower vs. leader, "regular" person vs. shaman, and Cofán vs. non-Cofán. Together, the oppositions express the polarized and paradoxical nature of Cofán society.

The difference between Antonio and Pablo involves the question of value, which we can investigate in two analytical moments.[2] First, a people's values are their key "conceptions of the desirable" (Kluckhohn 1951). As a set of concepts, metaphors, and standards, values allow individuals to make everyday choices, to devise overarching life projects, and to decide whether their current circumstances are satisfactory or unsatisfactory, and thus deserving of maintenance or transformation. Second, value (as opposed to "values") has a more quantitative sense as the enabling condition(s) of productive activity. As people pursue their visions of a desirable life, they produce themselves and their world as a set of material objects, embodied qualities, and social relations. Their possession, embodiment, and inhabitance of such objects, qualities, and relations enable them to act in the contexts that compose their lives. The more value to which a person has access—or the more advantageous their position in a matrix of social and natural conditions—the better their capacity to live the life they want.

Although my initial definition may appear confusing, I illustrate its meaning and utility in this chapter, which offers a value-based analysis of Cofán social relations and political perspectives. Cofán values shape how individuals in Zábalo and other communities conceptualize a desirable existence. The manner in which their current situation diverges from their ideals structures how they understand the necessity, meaning, and form of political action. Without a sufficient answer to the question of value, it is impossible to understand a collectivity's identity or aspirations. In the Cofán case, a value-based approach allows us to comprehend the underlying social and political dynamics of the Cofán experiment.

OPATSSI AND THE OPA A'I

In Zábalo, individuals judge themselves and each other to be Cofán if they practice preferred forms of sociality. They describe their social ideal with the word *opatssi* (the adjective form of the stem *opa*), which is the central term in Cofán value discourse. People use *opatssi* to characterize a good person, a happy community, and a satisfying existence. Upon being asked to explicate the term, my consultants turned to an animal metaphor: the *opa con'sin* (*opa* woolly monkey). Dwelling in areas far from human settlements, an *opa con'sin* has no idea what a hunter is. Without fear, embarrassment, or suspicion, an *opa* monkey wanders confidently through fruiting forests, happily gorging itself while oblivious to the specter of lurking enemies. Zábalo residents use the affect-laden image of the *opa con'sin* to express a convivial ideal of calm, satiation, and energetic health. People who enjoy and contribute to an *opatssi* state are *opa a'i*. Cofán oppose *opa a'i* to individuals, communities, and ethnic groups that are troubled by fear, tension, need, and violence. In large part, to be Cofán is to be desirous and capable of an *opatssi* existence.

Although Cofán employ the stem *opa* to denote a central characteristic of an actor (e.g., an *opa a'i* or an *opa con'sin*), the lexeme occurs more frequently as an adverb or adjective. When reflecting on the good life in Zábalo, my consultants declared, "*Opatsse* we live" and "This is an *opatssi* community." The word also figures in social axioms, such as, "Together as a group of Cofán, you will be *opatssi*." To explain the meaning of *opatssi*, people use a cluster of words with related senses: *dyombitssi* (not fearful), *ansa'ngembitssi* (not prone to timidity, nervousness, or embarrassment), *ppimppintssi* (calm like still water), *zíyatssi* (silent), and generally positive terms like *avujatssi* (happy) or *ñotssi* (good). One man explained:

> *Opatssi* means living without being embarrassed or anxious. It means that you'll be happy. You sit, you rest, you sleep, you're full. You go to your field, and you come back. You're satiated. You bathe in the river. It's like the *opa con'sin*. That's what we're like here. If you're afraid, you'll have many troubling thoughts. But if you're *opatssi*, well, like that woolly monkey, it goes high, high up in the trees. And it looks all around. It sees you on the ground, and it comes very close. It doesn't know. It doesn't know how to be afraid. It doesn't know how to be nervous or timid. It will come to you and stare at you. That's what an *opa con'sin* is like.

The typical imagined scene concerns animals that live far from human habitation and hunting, which make game *injansu* (suspicious): always on the lookout and always ready to flee. *Opaetsoñe* (literally, to make *opa*), however, is also a synonym for *aiñañe* (to raise an animal as a pet).[3] On the one hand, *opatssi* animals are those who are closest to humans, as residents of the same house, eaters of the same food, and participants in the same domestic scene. On the other hand, *opatssi* beings are those who are farthest from humans, inhabiting spaces where people rarely tread. Pets are similar to unhunted animals because they are products of an equalizing process, which removes their awareness of potentially violent difference. The word *aiñañe*, after all, combines *a'i* with a causative suffix. Accordingly, to make something *opatssi* is to make it an *a'i*, at least in relation to other *a'i*. Between beings that are *a'i*, there is no reason for fear, conflict, or suspicion.

The same holds true in the realm of everyday social relations. Cofán people prefer to live by Cofán norms in Cofán communities. It is best if everyone is socialized as an *a'i*, even if their genealogies betray their origins with other peoples. The Cofán believe that ethnic others are prone to stinginess, anger, violence, and killing, all of which are anathema to *opatssi* conviviality. One Zábalo resident compared the peace and quiet of his own community to neighboring Napo villages. He criticized Napo for their drunkenness, anger, and fighting, proclaiming, "We Cofán people, residing together as a group, would never suffer, struggle, and live like that."

Cofán people would like all of their coresidents to be *antia* (kin), if possible. *Antia* ties are more important in the paternal line, but one's mother's paternal relatives are also *antia*, to at least the degree of second cousins. Marriage with them is prohibited. In general, however, Cofán downplay affinity (in-law relations) and stress consanguinity (ties by "blood"). In A'ingae, there are very few affinal terms. The words that do exist—namely in the child-in-law, parent-in-law relations—operate mainly at the level of reference. Address terms are either consanguineal or slight transformations thereof.

A Zábalo family (Photo by Michael Cepek)

Extended coresidence shifts address forms. It is very common for children to spend significant time with either set of grandparents. They address them as "mother" or "father,"[4] and they address their genealogical "uncles" and "aunts" as "older brother" or "older sister."[5] Virtually all relations with younger individuals are covered by the terms *jaricho* (son, grandson, nephew, younger brother, younger male cousin, unrelated young man) and *sheno* (daughter, granddaughter, younger sister, niece, younger female cousin, unrelated young woman). Although memories of "exact" ties remain, terminological and behavioral fluidity bolster a generalized stance of affection and mutual care.

The Cofán borrowed godparent and "accompaniment-parenthood" (Uzendoski 2005:79) practices with relish. If people are not already *antia*, they use ritual kinship to make each other into relatives. The effect is a near-total web of relatedness in a community, which assumes the shape of a family writ large. All literal and fictive kin affirm their relations through the same forms of practical and material reciprocity. The prototypical expression of continuous, reciprocal relations is the sharing of meat. In one man's words:

> This is a happy community. We together, as a group, are happy. It's like we're all living together as *antia*. If you get a little bit of meat, you bring it over and give it to your friend. If you get a lot, you give it to someone else, to unrelated people. You say, "Here, eat this," and they become really happy. Those are good customs. We *a'i* know to be like this. That's what it's like here.

An *opa con'sin* is full and unworried in a forest of limitless fruit. Similarly, an *opatssi* community is characterized by abundant food and generalized reciprocity. Ideally, people neither keep track of past acts of giving and receiving nor do they worry about future necessities or the likelihood of return. The focus is on ease, cooperation, plenty, and generosity. The "knowledgeable giver" is the object of accolades, and the "stingy prohibitor" is the object of critique. If a person revels in their relative wealth, others will question their Cofán-ness. Although I witnessed few acts of aggressive boasting during my time in Zábalo, some community members told me that certain individuals insult their poorer coresidents. Their action is the object of negative judgment and the occasion for calls for equality. One man explained:

> Some people loudly brag to others. They'll say, "I have money. I have a motor. I have everything. You have nothing." They'll say, "That person is poor." That kind of talk exists here. But we still live quietly. If someone really

does begin to speak like that, other people will want to meet and talk so that everyone becomes equal, so that everyone has a little bit of money.

In addition to the positive compulsions to share and to cooperate, a set of negative injunctions structure life in Cofán communities: do not insult, do not fight, do not kill, and do not sleep with others' spouses. Libidinal desire and material inequality always exist as potential disturbances to *opatssi* life. Nevertheless, an *opa a'i* usually keeps them in check. During my time in Zábalo, I never saw an act of physical violence, domestic or otherwise. The explicit condemnation of such activities, however, shows that Cofán people view them as dangerous possibilities.

An ideal community is characterized by peace, quiet, and general good-will. Reflecting on such an existence, Cofán claim that to live *opatssi* is to be without anxious thought of any kind. Accordingly, an *opatssi* stance corre-lates with the Cofán preference for marrying within a community. A person does not want their children or parents to live far away. The distance of loved ones causes interminable fretting: Are they healthy? Are they hungry? Are they happy? There are no set rules for postmarital residence, but all parents try hard to keep their children close by.

Cofán people project the sense of community-as-kin onto the history of the entire Aguarico region. Decades of colonization, nation-state expansion, and petroleum-based development caused massive ecological and social alterations. The area is now home to many kinds of people, whose presence makes it impossible for contemporary Cofán to travel in an *opatssi* fashion. One can no longer fish, hunt, or gather wherever they want, moving between temporary camps and the communities of kin and friends. In many places, game is scarce. Moreover, pollutants from the oil industry and colonist agriculture threaten Cofán health. Before the arrival of the newcomers, the Cofán traveled through their lands *a'qquia injanga* (without structure, intent, or concern). Now, their territory is a mixed land of difficult and dangerous others. In one person's words, "In your own land, you already know everything, and you'll be *opatssi*. But if you go somewhere else, to someone else's land, you'll be afraid, afraid of doing something that the owner will prohibit. That's why it's difficult."

When Cofán people talk about living within an alien structure, they contrast it to acting "with one's own heart." The idea of an organic fit between structure and practice forms the experiential center of the *opa* ideal. If individuals have to discipline their behavior according to the rules, laws, or presence of others, they cannot live *opatssi*. Cofán people express their preferred mental life with the phrase *asi'ttaen'chove me'cho* (without troubling preoccupation or thought). Outside of their communities, they confront hostile others and anxious situations. Many past leaders abandoned political work because of the incessant traveling and cross-cultural interaction it demands, which lead to stress, worry, and excessive mental exertion.

At the most general level, a lighthearted irreverence forms the ideal tone of Cofán life. Zábalo sociality conforms to a common Amazonian norm, which Joanna Overing describes as a "looseness of structure" (1989:160). In Cofán society, social roles do not bind with the power of the holy, and anything can be laughed at. Cofán humor, which is constant and all-pervading in forms of *co'feye* (playing or joking) and *da'ñoñ'cho aya'fa* (damaged language), offers a mechanism to distance any sense of troubling structure or concern. People defuse anxious predicaments with ground-clearing bursts of self-effacing laughter. One of my favorite examples of Cofán people's humorous disregard for structure involved Antonio and Saura. When I asked them the names of their parents, they admitted that their mothers shared the same paternal *antia* name. Consequently, their union violated the taboo against marrying "together as *antia*." Instead of showing shame, however, they looked at me and laughed, proclaiming, "We are like kin!"

In general, Cofán people seek to minimize anxiety, whether it arises from explicit rule, social conflict, material need, bodily sickness, or political crisis. The desire to live a calm, peaceful, satiated, energetic, and joyful life—the *opa* ideal—becomes a salient goal in relation to circumstances that are of both recent and ancient origin. There was always hunger, contingency, and potential violence from neighboring peoples. Moreover, people always had to worry about the health of their kin and friends. Currently, however, a complex social, political, and economic situation magnifies Cofán concerns, even in the most intimate spaces of their lives.

It is important to stress, however, that the *opa* ideal does not refer to an ahistorical state of pure beginnings, "before the outsiders came." Cofán people now depend on Western medical care to treat life-threatening diseases. Moreover, the overwhelming majority of my informants stressed that they are prepared to sacrifice time and labor for access to a number of commodities (e.g., salt, soap, firearms, and outboard motors), which they now consider to be necessities. The point, however, is that the goods are tools that allow them to be more, rather than less, *opatssi*. Consequently, people are not willing to give up the joys of community life to secure limitless supplies of these commodities.

As the key Cofán conception of the desirable, the value of *opatssi* allows people to make decisions and evaluate circumstances on a number of scales. As they confront relatively open futures, Cofán people ask themselves and each other: How should I act toward my coresidents? In what places and with what kinds of people should I live? To what degree is paid labor worth the sacrifice of staying away from family and friends? What are the risks and benefits of sending children to outside schools? What kinds of leaders would best represent us? And, most importantly, how could we make our lives better?

The questions involve a discourse of community life rather than individual persona. They involve consideration of all the relations and practices that I described above, from generalized reciprocity to peaceful sociality to material plenty to frequent humor. In confronting current circumstances, Cofán people orient their choices around the ideal of maintaining the highest possible quotient of joyful, quiet, and anxiety-free *opatssi* conviviality. People's support for political projects depends on their understanding of how transformative efforts will allow them to be more, rather than less, *opatssi*.

I hope that my use of the first analytic sense of value—a key conception of the desirable—is clear. I fear, however, that my employment of the second sense of value—the enabling condition(s) of productive activity—may still be overly implicit. The basic issue is how, in his or her pursuit of an *opatssi* life, a person produces an ensemble of material objects, embodied qualities, and social relations that express and enable that life. Materially, *opatssi*-oriented action creates an overall condition of ease and plenty, especially in terms of food and other objects that make community life stable and satisfying. Psychologically and morally, *opatssi*-oriented action produces people who are capable of calm, goodwill, and generosity. Successfully socialized *opa a'i* devote themselves to energetic labor, generalized reciprocity, and leisurely consumption. And socially, *opatssi*-oriented action produces a group of people who inhabit a set of symmetrical relations. If a community attains the form of a family writ large, in which everyone acts as *antia* toward one another, dominance and hierarchy are undesirable and even impossible.

By striving for an *opatssi* life, people produce an *opatssi* existence for themselves and their coresidents. The pursuit of an *opatssi* life is not a zero-sum game. One's *opa*-ness increases the *opatssi* state of one's coresidents, and vice versa. *Opa a'i* are *ñotssi a'i* (good people), and they are the kinds of individuals whom one wants as friends, spouses, and neighbors. Even though there is a quantitative component to the value of *opa*, the logic is integrative and democratic rather than polarized and differential. The flexibility and minimal complexity of Cofán social structure—whether in terms of gender, generation, affinity, or interhousehold relations—constitute a social field in which one's satisfaction does not depend on the submission, subjugation, or exploitation of others.

If community members realize an *opatssi* state, they become equals, at least in terms of the value to which each has access. Nevertheless, the value of *opatssi* also has a contrastive function. Although it does not operate to differentiate coresidents from one another, it allows individuals to contrast their situation with all that exists outside of contemporary community life, whether other peoples, places, or times. *Opatssi*'s necessary contrast with an external realm of otherness makes it a utopian value. A truly satisfying socioexistential state is always precarious and partial. Through incomplete socialization of once different others, desire, need, and anger threaten a community's interior calm. Moreover, there are always external enemies, who wait to rip a community apart at its foundations. *Opa*, after all, refers to an ideal of self and community. The life it promises can never be fully realized.

The impossibility of a truly *opatssi* state accounts for the function of an important figure: the shaman. By examining the nature of Cofán shamanism, we can understand the ultimate paradox of Cofán politics: in order to enable a maximally *opatssi* life for the majority of Cofán people, some Cofán individuals must forego that life for themselves, thereby complicating the identity and values that join them to the wider Cofán nation.

COFÁN SHAMANISM

When I returned to Chicago after the end of my dissertation fieldwork in late 2002, I wrote an e-mail to Randy Borman about the word *opatssi*. Although I noticed the importance of the term while in Zábalo, it was not until I had a chance to read my notes and transcribe my interviews that I recognized its centrality. Finally, I thought that I understood the Cofán vision of the good life, and thus the foundation of Cofán identity and Cofán politics. Borman agreed with some aspects of my interpretation, but he said that I basically got *opatssi* wrong.

Borman's response surprised and troubled me. He asserted that *opa*'s main use is not to signal a politically motivating vision of an ideal existence. Rather, he claimed that people use the term to criticize individuals who are naïve and powerless. Furthermore, he suggested that the word occurs more in discussions of shamanism than in longing statements about a utopian past or a hopeful future. After rereading my notes and transcripts, I did find evidence that supported Borman's interpretation. To my surprise, I discovered that some Zábalo residents, such as Pablo, define an *opa a'i* negatively —as the meek and ignorant opposite of a powerful shaman.

The opposition, however, is far from simple. First, living in a state of "fearlessness" is an ideal of both figures. Second, the relationship between the *opa a'i* and the shaman forms the division of moral labor at the heart of Cofán communities. In short, *opa a'i* depend on the non-*opatssi* qualities of shamans in order to attain some version of the life that both seek: an existence in which neither is bound by fear or anxiety. Even with my newly won understanding, however, I remained troubled by one question: Why would Borman affirm one side of *opa*'s meaning (its negative sense from the perspective of a shaman) and deny the other (its positive sense for everyone else)?

As a point of departure, consider the following statements, which helped to expand my understanding of the *opatssi* ideal in Cofán society:

> "*Opa a'i.*" That means us, not a shaman. One who doesn't know, who isn't a *curaga*, who isn't a shaman.
>
> We *a'i* are *opa a'i*. We don't know. We don't understand.
>
> When one tries to drink [hallucinogens], becomes afraid, and stops, then we say that he is an *opa a'i*. Some people try drinking and go a little further. With their own eyes they see the *yaje* [supernatural] people. But they become afraid, and they then stop for good. That's how they live, how an *opa a'i* lives. Some people, though, see and want to drink more. They will keep drinking. They struggle and suffer.

As the above passages suggest, not all Cofán people are shamans, especially if they consider themselves to be *opa a'i*. Nevertheless, all Cofán people know something about shamanism. Typically, most individuals, including all men and many women, consume *yaje* (*Banisteriopsis caapi*) with various plant additives (*opirito* [*Psychotria pschotriaefolia*] and *yaje o'cco* [*Banisteriopsis rusbyana*]) during their youth. They drink the hallucinogenic mixture in a collective setting, at a special location (the *yaje tsa'o* [house]), and with the guidance of a knowledgeable elder who serves the brew.

People with sufficient drive go further, ingesting greater and greater quantities of *yaje* as well as other vision-producing substances, including *va'u* (*Brugmansia suaveolensis*), *ccumba* (tobacco, in various solid and liquid forms), *yoco* (*Paullina yoco*), *tsontimba'cco* (*Brunfelsia* sp.), and a few other plants whose scientific names I do not know. Once they have acquired power, a knowledgeable person becomes "one who has seen," "one who has learned," or "one who has transformed." The Cofán identify a particularly powerful person as a *curaga*. The word appears to be an A'ingae rendition of the Quechua term *kuraka*, which stood for an indigenous Andean leader who mediated between native followers and external sources of power, including deities and Spanish colonials (Stern 1993:41–44, 175; Szeminski 1987:183).

People explain their choice to progress on the shamanic path as the desire to see, to know, and to cure. Individuals who relent admit their inability to deal with the terror of the visionary experience, the suffering of initiation and ongoing prohibitions, and the moral ambivalence of the shamanic vocation, which inevitably involves killing. Eventually, through individual pursuit as well as guided learning and exchange, a shaman develops a number of powers. Typically, people separate shamanic capacities into five classes: the accumulation of relations of aid and familiarity with supernatural *a'i* or Cofán-like people; the power to see, speak with, command, and transform oneself into *cocoya* (malevolent supernatural agents); the acquisition of different kinds of *davu* (glass- and rocklike spirit projectiles); the ability to "call" and "make" fish and game animals; and the capacity to command or transform oneself into nonhuman predators, the most important of which is the jaguar.

As Borman suggested, and as my reassessment of the data confirmed, shamans do not lead *opatssi* lives. When describing individuals who fail to acquire or maintain shamanic power, people use the word *opa*. They say that *opa a'i* do not know, do not understand, and are too afraid. Making sense of their usage of *opa* requires a closer examination of three aspects of the shamanic path: its structure, its violence, and its mediation of threatening difference.

Whereas flexibility and disarming humor are key qualities of Cofán life, the shaman is the very embodiment of painfully binding structure. If a person wants to obtain shamanic power, he[6] must undergo an intense initiation period. His body becomes "hard" and cleansed of impurities through the daily consumption and vomiting of emetics. He eats almost no meat for periods of a year or more. He cannot consume any food or drink that was not prepared by himself or under his close supervision, for a menstruating woman or a pregnant person (either the prospective mother or father) may have contaminated it through touch or close proximity. He cannot cross the path of any contaminating individual. He cannot let a potentially unclean person walk behind his back or over any object (pot, cup, or gourd) from which he might eat or drink. Finally, he cannot have sex during either the initial learning period or any subsequent acquisition of powers or weapons.

When a person begins to drink *yaje*, the *yaje a'i* (a class of potentially friendly supernatural people, who themselves are shamans) give him a clean white spirit tunic as well as a headdress and ornamentation. If he breaks any of the prohibitions, the garments become red, smeared with blood. He loses his own blood and all of his strength. He gets horribly drunk on *yaje*, vomiting, urinating, and defecating all over himself. When he drinks, he sees nothing. All of the spirit objects he had acquired, which were like hard and sharp rocks, machetes, spears, and shotguns, kept inside his arms, depart from his body, causing serious harm. Supernatural people and *cocoya* become angry with and disgusted by him. A powerful shaman knows how to maintain the boundaries of his body better than a neophyte, and he can forego some of the prohibitions. Nevertheless, he, too, must remain vigilant.

The gendering of the shaman and the *opa a'i* is a complicated issue. Both men and women pursue the value of *opa*, and Cofán social life is relatively egalitarian: some women hunt; all men garden, cook, and clean; and domestic violence between Cofán spouses is rare. In contrast to the relatively ungendered *opa a'i*, the shaman is almost always male, even though exceptions do exist. My male and female consultants explained that women can and do drink *yaje*. Nevertheless, menstruation and pregnancy make their bodies unappealing to *cocoya* and *yaje a'i*, who tend to bestow their supernatural favors upon men. The gender imbalance places women more firmly in the domain of *opatssi* community life. Men, in contrast, have more opportunities to associate themselves with structure, violence, and otherness. Consequently, they dominate the shamanic realm.

Becoming a shaman entails forfeiting the hedonistic pleasures of the *opatssi* life. Gone are the satiation, the commensalism, and the easy wandering. Many of the most powerful shamans choose to cook, eat, and reside by themselves or possibly in the company of a postmenopausal woman. They do not trust anyone to follow the prohibitions and help them maintain their power and health. People sometimes refer to a strict shaman as an *asettu'cho a'i* (tabooing or prohibiting person), which is another antonym for *opa a'i*. Explaining the label, people say that one who desires power must "suffer and see." An aspiring shaman is constantly cautious about the relations between his body and the world. Attendance to rule becomes everything, on pain of sickness and death.

BOLD JAGUARS AND UNSUSPECTING MONKEYS

The collective valorization of the shaman is a strange and contradictory thing. All my Cofán consultants told me that a powerful shaman can cure, call game, and ensure a conflict-free social state. A shaman wins his positive powers, however, by identifying himself with all that is violent and morally condemned. If a shaman can cure a person of being "caught" by a *cocoya*, it means that he is a familiar of the malevolent being. He speaks its language and controls the entrance and exit to its house, where he can lock it up, thus effecting the cure. If a shaman can remove enemy *davu* from a person's body, it means that he has *davu* himself, which he acquired through friendly relations with other violent agents. To cure a person of *davu* is to extract the object and give it to a spirit ally, who eats it and thereby kills the enemy shaman or at least one of his close kin. If a shaman can call animals, it means that he knows the species' spirit master. Supernatural game owners, however, are always out for human blood. In certain cases, such as the "making" of fish, a shaman acquires the animals by promising a human meal to their anaconda owner. Only through a sleight of hand does the shaman appease the anaconda. He magically makes it see a deer or a paca as a human, thus feeding it a "person" to satisfy its hunger and cancel his debt.

In addition, a shaman can only solve real-world disputes because everyone is afraid of him. Shamans are prone to capricious killing. Stories abound of *curaga* who turn into jaguars and eat their coresidents. Many people believe that a shaman will shoot *davu* into the body of anyone who angers him. Although the benefits of living with a shaman are significant, people also disparage him behind his back. They call him "fool," "idiot," "demon," and "jaguar." People sometimes speak of shamans as irrational children who cannot control their emotions. Although the desire to do good for family and friends might have motivated their initial entrance onto the shamanic path, they inevitably take on the qualities of the malicious, foreign beings with whom they consort. The most powerful among them become *cocoya* after their death.

If a Cofán village is without a potentially violent shaman, social disputes can lead to complete fission. Numerous Zábalo residents told me about the long-dead *curaga* of Doreno, who enjoyed a fearsome reputation. In the late 1940s, a Shuar man worked for an oil company and fled to Doreno to avoid a vengeance killing. He married a Cofán woman and learned A'ingae. Unfortunately, his violent ways did not change. He beat his wife and threatened everyone who passed his house. The Cofán shaman told him to behave

A shaman adorned for a night of drinking *yaje*
(Photo by Michael Cepek)

himself, but he refused. Then, the shaman ordered other community members to apprehend and bind the newcomer. He whipped the Shuar man with tapir hide in front of the whole village, extracting an apology and a pledge never again to disrupt the collective *opatssi* state. Without the shaman's violent capacities, *opa a'i* never would have resolved the situation, and the community may have split into separate family groups. Ultimately, the shaman is the foundation of community integration, at least in the less-than-ideal circumstances that forever impinge upon an unstable *opatssi* state.

The lifeworld of a Cofán shaman is trying and terrifying. He foregoes all forms of pleasurable conviviality to prepare his body for interactions with the alien and the antagonistic. The violent otherworld, however, is not an ahistorical phantasmagoria. Many of the *cocoya* who inhabit tree trunks and foggy forest villages, which are only seen in dreams and visions, look like horrific versions of Gringo, Cocama, and Singo. They wear fine store-bought clothes or the black robes of colonial priests (Kohn 2002), even though they might be as tall as a kapok tree or armed with gigantic teeth. Cofán cosmology bears the strong imprint of violent colonial encounters.

On a more this-worldly plane, many shamans acquire spirit darts from Napo masters. Some learn to transform into a jaguar under the tutelage of Ai'pa experts. As in other areas of Amazonia (Carneiro da Cunha 1998), Cofán shamans are intercultural creatures par excellence, and many are multilingual. Cofán people interpret the social conflicts that a shaman negotiates as the eruption of non-Cofán violence in a Cofán community. And when an alien presence manifests itself in social disruption, Cofán people expect the shaman to deal with it in its own language: that of violence and killing. Any shaman worth his salt knows this language well, which is why *opa a'i* see shamans as morally questionable social others, even if they depend on them to ensure a desirable life. In opposition to the basic stance of a woolly monkey–like *opa a'i*, a shaman heeds a set of decidedly jaguaresque injunctions: suspect, attack, struggle, overcome, and expand your powers, even at the expense of other *a'i*.

The shaman's role in the mediation of violent otherness alludes to more timely anxieties as well. In Pablo's reflections on a shaman's never-ending vigilance, supernatural arsenal, and defensive network, we can see the mirror image of the present political climate of northeastern Ecuador:

> When you have *davu*, when there is *davu* in your body, you fall asleep and emerge in the *davu* house. That's where you stay. And in that house lives the master of *davu*. We call that master a *vajo* (demon). He owns the *davu*. You go to that master, you enter his house, and you wake up there when

you fall asleep. That's why when they come, hidden, to try to kill you, you already know. That house is made entirely of mirrors, mirrors everywhere. You stay inside of that, and when they come, the master tells you, "They will kill you." You have your shotgun there, and your automatic weapon. Your enemy has come, like a soldier, and he wants to shoot and kill you. They're like guerrillas, like the FARC. They come close. But you already see, and you kill them first. And you're saved. The *davu* owner will always warn you, saying, "They have come to kill you."

Pablo's image is the most telling expression of the opposition between a shaman and an *opa a'i*. Even during sleep, the shaman remains in a demon's fortress. Infinite mirrors multiply his suspicion, and he keeps his machine guns trained on ever-present enemies. In the case of Colombia and its violence, the shamanic inflection of political dangers is even more direct. Ecuadorian Cofán claim that the Cofán shamans who live near or in Colombia are the most powerful. They shout rather than speak. They walk in huge strides. And they obey all of the prohibitions. In Colombia, Cofán claim that shamans are the *na'su* of the guerrillas, whom they called Tsampi'su (People of the Forest), the same term that they use for a relatively benign and often cooperative supernatural people. In Colombia, the word *cocoya* refers not just to malevolent other-worldly agents, but also to right-wing paramilitaries, who invade Cofán villages and assassinate Cofán leaders.

Against the background of violent political clashes, we can see the opposition between the two paradigmatic forms of Cofán fearlessness: that of the *opa a'i*, whose fearlessness is equivalent to his unawareness of enemies; and that of the shaman, whose fearlessness is evident in his all-consuming desire to identify, attack, and overcome enemies. Ultimately, any overarching value orientation combines the moral and the existential in its construction of a universe that only allows for certain kinds of fulfillment. In the Cofán case, the cosmos constantly crashes into waking life in the form of violent difference. Facing a never-ending onslaught, Cofán people's ultimate hope is for the calm, satiation, and quiet of the *opatssi* state.

To maximize the *opatssi* quality of their lives, however, people must strike a deal with the shaman, who takes the powers of violence, chaos, and death into himself, thereby both becoming and mediating them. Accordingly, an *opa a'i*'s strong hope is that the perduring, though always ambivalent, links of humanity, kinship, and friendship with the shaman will ease what would otherwise be unceasing attack. The division of labor between the shaman and the *opa a'i* combines the operations of myth and history, social structure and cosmology, daily routine and political project.

The desirability of an *opatssi* life takes shape against a backdrop of violent difference. The latter is not simply a cosmological concern. Rather, it is an aspect of everyday interactions with ethnic others as well as a history of calamitous confrontations with colonial and nation-state violence. Cofán people can only produce an *opatssi* life with the help of a shaman, who becomes the very embodiment of the structure, power, and difference that they want to evacuate from their lives. Accordingly, a Cofán community is a complex and compromised thing. It can only exist when a shaman occupies its center and periphery. Shamans act to integrate constantly disrupted collectivities by participating in the very source of their disruption. Consequently, they lose the basic identity—*opa*-ness as Cofán-ness—that ties them to their coresidents. As Eduardo Viveiros de Castro writes of the Araweté, Cofán society's center is outside of itself (1992:3).

SHAMANS, LEADERS, AND POLITICAL ACTION

Although their work occurs in such spaces as NGO offices and government ministries, contemporary Cofán leaders are similar to shamans in important ways. Historically, Cofán political action depended on an array of leader figures. All of them, however, exhibited the central characteristic of aggressive fearlessness in the face of external hostility. Although Cofán *curaga* continue to negotiate a violent otherworld, shamanism is no longer the most important institution for managing the crises that external conjunctures bring.

Like other indigenous peoples of lowland South America, the Cofán have had to organize themselves into state-recognized communities in order to gain legal control of their land. As political institutions, recognized communities must have a juro-political structure that includes a president, a vice president, a secretary, and a treasurer. Cofán people refer to holders of the first and second positions as *na'su*.[7] In addition, they classify all officers of their ethnic federations and political organizations as *na'su*. *Na'su* is the same term they use to refer to the powerful *curaga* of previous generations.

Although Cofán shamans continue their curative practices, community presidents and federation leaders now assume the majority of their historical responsibilities. In earlier eras, shamans were essential political-economic intermediaries. When threats came invisibly or from a great distance, shamans were crucial defenders of *opatssi* life. If a village was under attack from an enemy shaman's projectiles or *cocoya* familiars, or "commanded" jaguars and anacondas, only a friendly shaman could ensure security and stability. Shamans served a more economic function as well. During bouts of scarcity, they provided fish and game through supernatural means. In addition, their courage, aggressiveness, and cross-cultural facility made them preferred negotiators in relations with traders, priests, and government officials.

A different means of negotiating difference preceded both shamanic and "presidential" forms of leadership. Through the 1600s, the Cofán enjoyed a reputation as fierce warriors. They fought wars against enemy indigenous peoples in pre-Conquest times, and they repelled incursions by Incaic and Spanish forces. With a much larger population, the old-time Cofán were a tremendous military power. They were notorious for martyring the Jesuit priest Padre Rafael Ferrer in 1611 (Cabodevilla 1996:94–98), burning the Colombian town of Mocoa to the ground, and laying siege to the city of Pasto (Friede 1952).

A myth that most Cofán know well tells the exploits of "Erición," a Cofán boy who was kidnapped, raised, and trained in warfare by Tetete. On the day that he was to lead Tetete warriors in an attack on a Cofán village, Erición tricked his adoptive nation and reclaimed his Cofán identity. After ornamenting himself with a tunic and feather crown, he turned on his captors and used his incredible powers to exterminate all of the Tetete and Ai'pa who resided between the upper Aguarico, the middle Napo, and the main body of the Amazon. When I asked young Cofán men what they would do if Colombian paramilitaries invaded their Ecuadorian villages, they joked about reclaiming Erición's powers. Similar to a shaman, Erición consumed and vomited huge quantities of powerful plants in order to acquire immense strength and a rock-hard physique. Such intense training produces a body that neither enemy spears nor enemy bullets can penetrate.

From the perspective of a laughing and peaceful *opa a'i*, old-time Cofán appear as true brutes. My Zábalo consultants claimed that their ancestors were "bad people" and "killers." When contemporary Cofán admit the warrior violence of their forebears, they ethnically distance them. Old-time Cofán appear to have appreciated none of the *opatssi* qualities that form the core of contemporary Cofán society and identity. While expressing his fears of Cocama life in Lago Agrio and Colombia, a Zábalo resident said, "We *a'i* have never killed each other like that." I protested, arguing that the man himself told me many stories about the violent ways of old-time Cofán. He retorted:

> Yes, that's how the old-time Cofán lived. They thought and desired as Tetete. That's why as soon as they became angry, they would just kill the person they were mad at and throw them away like garbage. We say that they were bad people, those old-time Cofán. Now, we've changed. We have other ways of thinking and desiring.

Although their political position is still novel, contemporary leaders bear a resemblance to shamans and warriors. They enter enthusiastically into tense and difficult relations with powerful outsiders. They speak different languages. And Cofán people expect them to fulfill both defensive and acquisitive functions. They work to protect Cofán territory, and they seek out sources of desired goods and services. Although their intercultural work occurs in neither hallucinogenic visions nor bloody battlefields, it continues to articulate with the power and violence of ethnic difference.

Currently, there is a cultural and moral gap between Cofán people and the people who lead them. Most Cofán are loath to decrease the *opatssi* quality of their lives for any reason. Even though community members elect their leaders to office,[8] they are suspicious of anyone who dwells in alien contexts and consorts with violent others. Whereas *opa a'i* expect reciprocity and generosity within a community, they do not assume the goodwill of actors who claim to engage in extra-community affairs for altruistic reasons. For the majority of Cofán people, acting on behalf of "the Cofán nation" remains an abstract and questionable motivation.

Their assumed ambivalence makes all Cofán leaders easy targets for accusations of corruption or womanizing. Many find it so difficult to deal with the insults and "bad talk" that they give up on political work altogether. Nevertheless, all Cofán people know that avoiding outside forces is no longer an option. They realize that they must accept the morally ambivalent actions of their new *na'su* if they are to maintain any quotient of *opatssi* conviviality. After all, the effectiveness of leaders depends on their difference from other Cofán. Although Cofán question and criticize their *na'su*, they also depend on them, even though they would never want to become them.

Although they share the fearlessness of the shaman and the warrior, most contemporary *na'su* do not share their aggressiveness. For Borman, Cofán people's lack of ferocity is a political liability. When he alerted me to the double meaning of *opa*, he was in the process of writing a grant application to the MacArthur Foundation. He wanted to obtain funds to train and equip a corps of Cofán park guards. In his proposal, he portrayed the group as a legalized police force with the power to confront coca-growing colonists, to make boundary trails in FARC-controlled land, and to face off against oil and mining companies.[9] *Opa a'i*, of course, would be hesitant to engage in any of these activities, as Borman knows well.

Borman has been working for decades to reinstill a warrior sensibility into what he understands as a far-too-*opa* Cofán nation. Although no shaman, he identifies with a jaguar. He placed a "jaguar curse" on an invading Secoya community in order to scare off the newcomers. He sometimes wears a tooth of the animal around his neck. And he paints whiskers along his nose for political meetings and interviews with foreign journalists. It should come as no surprise that his assessment of the value of *opa* matched neither my own nor the greater Cofán population's understanding.

By negatively characterizing *opa a'i*, Borman questioned the value of Cofán values. From his perspective, the Cofán nation needs battalions of face-painted and spear-wielding warriors if it is to arouse the fear of Ecuador's mestizo population as well as the sympathy of the international media, which loves to romanticize indigenous struggles. In meetings with colonist organizations and government officials, Borman sometimes invokes the specter of "one thousand Cofán warriors." Given the *opatssi* state of the contemporary Cofán nation, however, his threat does not (yet) correspond to reality.

The trajectory of shifts in leadership positions, however, is not one-way. I know of one resident of an upriver Cofán settlement who served as an elected officer of his own community and the Cofán federation. His work took him to the United States and Europe, and he traveled constantly through Ecuador. As a political activist, he was humble, kind, and free of corruption. Nevertheless, he lacked the requisite fire and charisma. In a word, he was too *opa*. In 2004, I spoke to a friend from the leader's home village. He told me that the man was working as a community schoolteacher and drinking *yaje* with a knowledgeable elder in order to accumulate shamanic powers. Knowing how much time the leader spent in social spaces where he could not obey the required prohibitions, I asked how he could become a shaman. My friend replied, "You knew the man years ago. When's the last time you saw him in a town, eating at a restaurant?"

If my friend's account is correct, the leader has given up one kind of mobility for another. He appears to be shedding his *opatssi* stance in order to embrace a new kind of fearlessness. Whether he will be able to use a different set of dispositions in renewed political engagements is an open question. Nevertheless, his decision to embark on the shamanic path demonstrates that the figures of the president, the warrior, and the shaman remain attractive and intertwined options for Cofán people who desire something more than an *opatssi* life, at least for themselves.

ANALYZING INDIGENOUS POLITICS

Analyzing the value of *opa* speaks to many of the main concerns of anthropologists who study contemporary indigenous movements. Kay Warren and Jean Jackson summarize some of the most important questions, including, "Who is entitled to represent indigenous populations? Who makes the decisions about who is entitled? What are the various ways in which indigenous peoples are represented in their own and in others' political imaginaries? What conflicts arise in the course of fashioning such representations?" (2002:27).

As the Cofán case makes clear, the answers are far from straightforward. Effective representation can operate through difference, contradiction, and distrust. Moreover, there are conflicting but interdependent visions of the political imaginary that forms the subject and object of Cofán mobilization. In the Cofán case, there is no homogeneous group dedicated to a simple, common goal. Instead, a polarized and paradoxical dynamic is at play. The political aspirations of most Cofán people revolve around increasing the *opatssi* quality of their lives. The *opa* ideal, however, acts as an impediment to collective mobilization because it affirms community-internal conviviality and rejects community-external confrontation.

Cofán people consider their leaders to be culturally different, morally ambivalent, and problematically oriented to alien actors, settings, and satisfactions. Their characteristics point to definitive patterns in Cofán social life, moral philosophy, and political practice. Considering the relationship between the *opa a'i* and the shaman, as well as the warrior and the contemporary leader, allows us to revise important assumptions regarding Amazonian politics. Many anthropologists question the status of the "new leaders" of Amazonian social movements (Brown 1993; Conklin 2002; Conklin and Graham 1995). They suggest that "chameleon-like leader[s] who can successfully walk in two worlds" are replacing "headmen" who work as "models of competency, generosity, and tact" (Brown 1993:310).

The shaman, the warrior, and the contemporary Cofán leader teach us that the new leaders might not be so new. Moreover, the array of Cofán *na'su* suggests that the logic of Amazonian politics need not involve identity, comprehensibility, or moral agreement between leader and led. Clearly, some representatives are more effective and appreciated than others. Nevertheless, observers and advocates of indigenous political struggles should realize that the dynamics are not as simple as they imagine. In their analyses, criticisms, and forms of support, they should be willing to entertain extremely complex scenarios.

The relation between community members who desire an *opatssi* life and the ambivalent individuals who act as their *na'su* plays an important part in the situations, activities, and projects that figure in the remainder of the book. In the most basic terms, Cofán political mobilization involves a great irony: the more individuals identify with their Cofán-ness (i.e., their *opa*-ness), the less open they are to direct participation in the struggle for their valued existence. Whether working in an urban NGO, participating in scientific conservation projects, or sending their children to outside schools, Cofán people make difficult sacrifices. The great majority are willing to leave the hard life of being a *na'su* to whoever will take the job, which is why the work of such ambiguous figures as Randy Borman is essential to their future.

II AN EXPERIMENT IN INDIGENOUS AND ENVIRONMENTAL POLITICS

4   *The NGO*

I began this book by introducing the three meanings of "possibility" that orient my approach to Cofán politics. The first sense—the conditional nature of Cofán institutions and accomplishments—is impossible to ignore when considering the Foundation for the Survival of the Cofán People (FSC). The organization was little more than a year old when I began my dissertation fieldwork in January of 2001. During my initial stay in Ecuador, the FSC was in perpetual crisis. A lack of funds and donor support produced an air of anxiety at the Quito Cofán Center (QCC). During my visit in July of 2010, tension was at an all-time high. Alarmed by questions concerning accounting, the Ecuadorian tax service was conducting an audit. Government officials literally blocked entry to FSC offices with thick strips of tape, which could not be broken until the investigation was over. Even though the FSC was paralyzed, its employees hoped that the problems would pass. No matter how intractable the difficulties, they always had been overcome before.

Despite the familiar sense of crisis, much had changed between 2001 and 2010. From a tiny enterprise with an annual budget of approximately $36,000, the FSC grew to demand more than a million dollars of yearly expenditures by the end of the decade. Although Borman and other FSC employees are uncertain about the institution's future, its accomplishments are impressive. With the FSC's financial and logistical aid, Cofán territory tripled in size. The FSC created a group of sixty Cofán park guards to patrol the land, oversee management activities, and prevent intrusions from colonists and corporations. In communities such as Zábalo, the FSC coordinates scientific conservation projects, which provide salaries and training to dozens of families. Perhaps most importantly, the FSC supports the education of Cofán youth in Quito schools, where they are working to attain the capacities that will make them future leaders of their people.

In this chapter, I analyze the history, operations, and vision of the FSC, which is the central institutional support for Cofán experiments in indigenous and environmental politics. My main observations pertain to the organization's early work in 2001 and 2002. An investigation of the FSC's first years has wide-ranging implications. As Cofán activists and their allies struggled to negotiate the requirements of a novel organizational form, they confronted the obstacles that face all peoples who make NGOs essential agents of their political projects.

In their work with the FSC, Cofán activists encounter, adopt, critique, reform, and resist the requirements of contemporary political practice: urban life, grant proposal formalization, bureaucratic interaction with ethnic others, and demands for accountability and transparency. The difficulties represent some of the main challenges confronting indigenous activists in the twenty-first century. Analyzing the problems as well as Cofán responses to them provides the material for a more general evaluation of the means by which outside institutions fund, support, and complicate subaltern struggles.

HISTORY, PROGRAMS, INDIVIDUALS

Like other indigenous Amazonians, the Cofán responded to the colonization of their territory by forming an ethnic federation in the latter half of the 1980s. By 1990, Cofán leaders had mobilized as the Association of Indigenous Communities of the Cofán Nationality (ACOINCO). Within five years, they renamed their federation the Indigenous Organization of the Cofán Nationality of Ecuador (OINCE). Neither ACOINCO nor OINCE, however, was fully functional. Both lacked a legal status with Ecuador's Ministry of Social Welfare. Without their "papers," they could not request funding from state or nonstate entities. According to Cofán leaders, the representatives of the much more populous Shuar- and Quichua-speaking peoples kept them in a client role within the national indigenous movement. Only in June of 2001 did the federation achieve legal personhood as the Indigenous Federation of the Cofán Nationality of Ecuador (FEINCE).

In biannual general assemblies, Ecuadorian Cofán elect FEINCE leaders, who are entitled to speak for the Cofán nation in relations with state, corporate, and civil nonprofit actors. FEINCE contains ten elected posts: President, Vice President, Treasurer, Secretary, Director of Education, Director of Community Development, Director of Health, Director of Women, Director of Youth, and Director of Territory. Although FEINCE has strengthened since 2001, many Cofán question its capacity for autonomous action. For years, the federation suffered a severe shortage of funds for transportation, an office, or collective action of any kind.

The struggles of FEINCE and its predecessors provided the immediate rationale for the creation of the FSC. The mid-1990s marked an important moment in Cofán mobilization. Randy Borman was OINCE president from 1994 to 1996. At the same time, he decided to enroll his children in Quito's Alliance Academy. He already had a small home in the capital, which he used to visit his extended family and to negotiate the urban side of Zábalo's tourism business. By the end of 2000, a border war with Peru and Colombian unrest intensified the violence in northeastern Ecuador. The tourism industry was decimated, which meant that Borman and other Cofán people lost their most importance source of income. When I began dissertation research in 2001, Borman spent little time in Zábalo, apart from visiting, hunting, and fishing trips during the children's school vacations. Politically and economically, Quito became his home.

Borman's move to Quito began his struggle to create the FSC as a new agent of Cofán politics. On grant applications, Borman writes that the FSC is "the technical branch of the Cofán nation," which works in conjunction with FEINCE, "the political and legal representative of the Cofán nation." In 2000, Borman worked to incorporate the FSC in Ecuador. Two years earlier, he filed paperwork to incorporate the Cofán Survival Fund (CSF) in the United States. The CSF manages U.S. fund-raising, and the FSC utilizes the funds for projects in Ecuador. Although the CSF is a legal entity, it has no employees. As a 501(c)(3) organization, it channels the money it receives to the FSC. Borman relies on a handful of U.S. volunteers to direct its operations.[1] Clark Vaughn, an old friend of Borman from Quito's missionary community, used his skills as a lawyer to help found both organizations. He also introduced Borman to Chip Harlow, a North American who donated $100,000 to buy the Quito property that became the QCC.

To date, the FSC has coordinated an impressive array of initiatives: a Cofán-staffed minifactory of fiberglass Ecocanoes; paid biological monitoring in Cofán communities; educational opportunities for Cofán students in Quito; training workshops for adults, including Spanish instruction for women; and a series of legal initiatives and political actions that developed into the Cofán nation's effective control of more than 430,000 hectares of its traditional territory. In 2003, the FSC created the Cofán Ranger Program. Currently, sixty technically trained, legally empowered, and regularly paid Cofán guards make boundary trails, evict colonists and loggers, prevent illegal intrusions from mining and oil companies, perform monitoring work, and help Cofán communities establish management structures. In 2009, the FSC constructed a field office in Lago Agrio, from which it coordinates the ranger program. Four years earlier, Borman established a related NGO—the Institute for Conservation and Environmental Training (ICCA)—in Quito. ICCA trains the park guards and interested outsiders in the protection, monitoring, and management techniques that have become the foundation of Cofán conservation efforts.[2]

An assortment of institutions fund FSC activities. The most important are the Field Museum of Natural History, the John D. and Catherine T.

MacArthur Foundation, the European Community, the United States Agency for International Development (USAID), the Nature Conservancy, the Gordon and Betty Moore Foundation, Conservation International, the Blue Moon Fund, the United Nations, the Disney Wildlife Conservation Fund, and the Edwards Mother Earth Foundation. The FSC is also working with the Ecuadorian state to become part of the new "Socio Bosque" program, which directs public money to individuals and communities who commit to conserving their forests. Through Conservation International's Global Conservation Fund, the FSC has secured a promise to match up to one million dollars in donations for a trust. If realized, the fund will cover many of the FSC's ongoing expenses.

At the end of 2002, the FSC employed five staff members: Randy Borman (as Executive Director), Roberto Aguinda (a Cofán man from Zábalo, as Technical Director), Amelia Quenamá (Borman's wife, as Logistical Director), Maria Luisa Lopez (a mestiza woman from Quito, as a licensed accountant), and Freddy Espinosa (Lopez's *quiteño* husband, as Director of Communications). The five employees work out of the QCC, which holds the main office of the FSC, the urban residence of the Borman family, and another building that houses Cofán students and visitors. A few years after I left the field, the FSC purchased an adjoining parcel of land, which allowed for increased office space and dormitory and meeting rooms for political events and ICCA courses.

As funding waxed and waned, a number of other employees joined and left the organization. They included additional accountants, project coordinators, administrative assistants, house parents for the Cofán students, and institutional development specialists. In recent grant applications, Borman argues the necessity of hiring an "environmental services specialist." If funded, the new employee will work to position the FSC and its programs within the emerging markets in carbon offsets, watershed preservation, biodiversity maintenance, and other "conservation products."

Although the FSC's staff has swollen at times to include nearly twenty employees, Borman, Quenamá, Aguinda, Espinosa, and Lopez are the only constants. As Executive Director, Borman is the true head of the FSC. He attends most of the meetings, makes most of the important decisions, writes most of the grant applications, and conceptualizes most of the strategies and projects. As Logistical Director, Quenamá is essential for coordinating many of the FSC's field activities. She determines the type and quantity of food and equipment needed for such tasks as land demarcation. The Technical Director, Aguinda, does most of the on-the-ground work in the lowlands. He constantly moves between communities, speaks with officials in Lago Agrio, and helps to transport people and equipment between Quito and Ecuador's

Amazonian region. Director of Communications Espinosa does most of the legal legwork in Quito. He continuously visits officials at government ministries to facilitate the completion of *trámites* (bureaucratic procedures and tasks). As a certified accountant, Lopez manages FSC finances. She receives funds and disburses money, making sure to keep an exhaustive record of all transactions.[3]

POLITICAL VISION

Our mission is to preserve our culture by protecting our environment: the rain forest that is our home. We realize our mission through: *Protection*: the legal acquisition and defense of our ancestral territories from outside interests. *Conservation*: the documentation, analysis, and preservation of the ecology of our rain forest environment. *Management*: the sustainable use of our biological and cultural resources to generate stable and environmentally friendly sources of income for Cofán communities. *Education*: the development among our youngest generations of the knowledge and skills needed to preserve our land base and to deal effectively with the outside world.

The above statement comes from a recent bilingual FSC pamphlet. In the introductions to many grant applications, Borman explains the environmental orientation of the FSC's mission with versions of the following paragraph, which describes how Cofán society and politics are grounded in the Amazonian forest:

Absolutely central to Cofán culture and identity is our relationship with what we call, simply, the Forest. Without our Forest, which has been the one constant throughout our history, we are no longer Cofán. All of the activities that we consider to be essential to our way of life—hunting, fishing, collecting forest products, and residing in our riverside villages—are made possible by the ecology of the Amazonian environment. Over half of our language describes the Forest; we cannot even speak without it. Various aspects of our relationship with the Forest have been widely recognized: our skills in medicinal plants, our knowledge of the Amazonian environment, and, more recently, our innovative approaches to rain forest conservation.

FSC materials carefully note that supporting Cofán conservation is not about "doing good" for history's victims. Borman does not want outsiders to see the Cofán as objects of aid. Rather, he portrays the Cofán nation as an essential partner in a global movement for environmental conservation. In another grant draft, he explains the FSC's position:

[P]art of my philosophy as President of FSC . . . is that there needs to be a recognition, on the part of the world, that what we are trying to accomplish as indigenous peoples in gaining land titles to our ancestral territories is ultimately to the benefit of the whole world. We are the front lines, and as such receive both the brunt of the conflict and the lion's share of the medals. But we are only the front lines, and the war we wage is not possible unless we have solid backing from our citizens—citizens of Earth: all who care whether the world survives into the next century or not.

Borman is a strong critic of conventional aid paradigms. When given the chance, he tries to convince foundation officials that their interventions are based on a mistaken logic. Most funding agencies intend to make the objects of their aid "self-sufficient." Moreover, they want to create projects that are "self-sustaining." They assume that aid is about unilateral charity rather than mutual interest. Their position has two strategic effects: first, it masks the benefits that northern countries derive from southern conservation; and second, it construes aid as a short-term, one-way relationship.

One afternoon in Quito, I listened as Borman argued his position in a meeting with an American consultant. The man's firm was competing for the right to manage over $6,000,000 in USAID funds, which were earmarked for conservation projects with indigenous peoples in Amazonian Ecuador. The acronym for the program was "CAIMAN," which stood for "Conservation of Managed Indigenous Areas." The consultant began by outlining the objective of the aid. He said that USAID wanted to promote conservation by supporting the management efforts of indigenous peoples with an "integrated sense of territory."

After listening to the consultant, Borman began to criticize the idea that indigenous people are the sole beneficiaries of the work and thus undeserving of payment for their participation. The consultant responded with a series of questions: Who would benefit from the projects? Why should someone be paid to conserve their own resources? Why should people demand money when no one is profiting? Would payment do more harm than good by corrupting indigenous ways of life? And given the fact that the donor community will not always be there to support "local efforts," what could be more important than self-sufficiency?

Borman responded by asserting that donor organizations must abandon the idea that the "local community is the primary beneficiary of conservation." He stated that the Cofán are working hard to protect "biodiversity that the world needs." The mere fact of the consultant's presence in Quito, as well as the large sum of money that his firm wanted to manage, proves that Cofán territory is of global interest. Borman argued that the situation was particularly irrational when half of one midlevel expatriate NGO salary was enough to cover the administrative expenses of the FSC for a year ($36,000 at the time). A better allotment of funds could finance indigenous efforts on a long-term, rather than a short-term, basis. In addition, Borman emphasized the fact that the world is a political-economic whole. He emphasized that everyone, including the Cofán, needs commodities that only money can buy. Cofán people, he asserted, are right to demand payment for protecting what the world wants and needs.

Borman ridiculed the consultant's objective of promoting self-sufficiency for local communities. He offered the example of ecotourism. After developing their community business for two decades, the people of Zábalo watched it disappear overnight when the U.S. State Department issued a travel warning on Cofán territory after "Señor Clinton" signed Plan Colombia. A potential visitor sent him an e-mail that he received from the U.S. Embassy. It stated that the "community of Cofan," just a few miles from Lago Agrio and the Colombian border, was "very dangerous." (There is no community called "Cofan," and Zábalo is more than one hundred miles from Lago Agrio.)

Borman argued that it is impossible to consider the Cofán apart from the vicissitudes of global political-economic realities. He suggested that if the world values Amazonian biodiversity and "ecological services," it must enter into a permanent partnership with Cofán people to protect, to conserve, and to manage their territory. Given the prevailing aid structure, the distribution of funds offers minimal benefits to local people. According to the consultant himself, "sophisticated outside NGOs" would absorb at least half of the USAID money. "Large national NGOs" would claim half of the remainder. In the best-case scenario, local project-implementing organizations such as the FSC would receive only one quarter of the funds. Nevertheless, Ecuador's system of corruption and nepotism would channel a large portion of the money to ineffective organizations with little experience and few accomplishments.

The consultant listened to Borman and appeared to agree with many of his points. Nevertheless, he claimed to be a realist, saying that he could not change the structure of international aid. Moreover, he admitted that his own livelihood was part of the system that Borman criticized. After all, he did not want to argue his way out of a job. I sensed that he wanted to delve deeper into the debate. Nevertheless, he did not push too hard. He knew that Borman could share his opinions with USAID officials and threaten the consultant's chances of securing the contract. (Ultimately, another company won the right to administer the CAIMAN grant.)

Borman's objective as FSC head is to have the Cofán claim the positions of better paid and more securely employed outsiders. He wants to integrate Cofán people into global scientific fields as researchers who collect data, perform analyses, and receive the same recognition as their Western counterparts. He envisions the Cofán as authorized executors of national conservation objectives on lands they control as ecologically protected ancestral territory. And he pushes donor organizations to abandon the paradigm of hit-and-run project assistance with the unrealistic goal of local self-sufficiency. Instead, he wants them to construct permanent relationships with Cofán people as partners who can protect and investigate valued ecosystems not just for themselves but for the whole world.

In the most general terms, Borman is struggling to create a world in which Cofán people escape their position as the unpaid and uncredited objects of scientific expertise, state power, and transnational conservation work to become their fully empowered agents. In other words, he wants the Cofán to become international scientists, state enforcement officers, and global conservation practitioners. To date, Cofán people have made great strides toward realizing Borman's vision, which they increasingly embrace as their own.

COFÁN ENVIRONMENTAL POLITICS

As a recognized Cofán representative, Borman depends on the support of other Cofán leaders and the greater Cofán nation. The FSC's projects are part of an encompassing movement of Cofán environmental politics. In November of 2002, Cofán leaders gathered at the FSC's Quito offices to discuss the overarching vision of Cofán conservationism. As the head of the FSC and FEINCE's Director of Territory, Borman called them to the capital to discuss Cofán strategy. The state had just accepted Cofán demands and designated 55,000 hectares of Amazonian rain forest and Andean foothills, including four Cofán communities, as the Cofán-Bermejo Ecological Reserve (RECB). According to the reserve declaration, the RECB is both ancestral Cofán territory and part of Ecuador's National System of Protected Areas. By giving the Cofán the power to administer and manage the reserve, the state made them legally empowered functionaries of the Ministry of Environment.

Most of the leaders who attended the 2002 meeting had more experience mediating community conflicts than debating state law. Aware of the disparity in knowledge, Borman began the event with a general A'ingae discussion on territorial issues. As a director of the FSC and FEINCE, he identified his central mission as "fighting for land." He connected his current position to a long history of working for community titles, beginning in Doreno in the mid-1970s and continuing in Zábalo from the mid-1980s. To the leaders who were less familiar with his ongoing activities, Borman explained that he now spends most of his time in Quito, where he works with the government and NGOs to increase the territorial holdings of the Cofán nation. He followed his introduction with a question: "Is land a necessity?"

A young couple from the border community of Chandia Na'e responded first. They said that land is important, and that their children and children's children must continue to know the forest and the life it provides. The president of the community of Sinangoe, who grew up in heavily colonized Dovuno, added that more colonists arrive every year. He stated that his people do not want to live on tiny, individual landholdings. "We must strongly secure our land," he proclaimed. Another Sinangoe leader chimed in. He stressed that the Cofán must expand, rather than simply maintain, their territory.

Some of the attendees had heard of the RECB prior to the meeting, but few had imagined the reserve's implications for the future of the Cofán nation. Borman asserted that the RECB is a promising legal precedent because it combines the categories of "Cofán land" and "government land." As it is "land that must be cared for," he explained, Cofán people regulate their activities under the guidelines of Ecuador's legal system for conservation areas. Borman stated that Cofán park guards will uphold the system and work as police to prevent illegal, environmentally destructive activities. The Cofán leaders were enthusiastic about the new role. Can they detain people and make arrests? Yes, said Borman, they will have the full backing of the Ecuadorian state. The leaders agreed that the new power was important. It will help them avoid the destruction of their lands, they concluded, even if it means placing an additional layer of environmental restrictions on Cofán people, including themselves.

Two middle-aged Zábalo men addressed the question of threats to Cofán territory. They asserted that their community had done much to protect its land. Playing on a word uttered by the Sinangoe representative (ta'etssi, literally "hard"), however, they said that they were worried that other Cofán communities might begin to "soften" on territorial protection. No matter in which community it occurs, they stated, the loss of land to colonists, oil companies, and miners is a direct concern to all Cofán. In a 1998 general vote, Cofán people decided that each community's land is collective patrimony, thereby making individual and community action the business of the entire Cofán nation.

Other attendees began to discuss the issue. Most of them agreed that corporate extraction and uncontrolled colonization pose the greatest dangers to Cofán territory. Others suggested that Cofán people themselves might threaten the ecological integrity of the Amazonian landscape. One man mentioned the possibility of cultural change among Cofán youth. Another worried that more remote communities might be politically inexperienced and susceptible to deceptive corporate promises. Others noted the possibility that untrustworthy individuals might become commercially indebted or involved in illicit Colombian dealings. All the leaders understood that a legalized park guard force could act to combat the problems.

Acknowledging the possibilities, Borman returned to the question of the RECB and whether the assembled representatives agreed with his plan to bring more territory under the protection of both the Cofán nation and the Ministry of Environment. Should he continue to work, he asked, to declare even more Cofán land as national protected areas and to establish more alliances with the state and conservationist NGOs? The others replied with grunted assents. One man added, "Fight! Your thoughts and desires are good." Borman accepted the affirmation. He told the leaders that he did not want to struggle toward a vision that other Cofán people did not understand or support.

The discussion ended on an optimistic note. Borman stated that his main desire was to have Cofán people living on their lands and speaking A'ingae for at least the next five hundred years. A leader from a community close to Colombia echoed Borman's objective. He lamented that Cofán people on the other side of the border have been forced to become Cocama because their land has been alienated or destroyed. A Zábalo resident pondered the possibilities. If the Cofán can use the RECB precedent to protect their land, he wagered, his children's children—twenty times over—will continue to be Cofán. He smiled and concluded, saying softly, "They will exist."

Unlike other indigenous landholdings, the RECB is unique in that the Ministry of Environment, rather than other state agencies, classified the area as ancestral Cofán territory. Additionally, the RECB is a special case in that the state recognizes its status as Cofán territory only insofar as Cofán people manage it according to the laws governing Ecuador's National System of Protected Areas. The Cofán nation, in other words, must balance its sovereign rights over the land (as ancestral indigenous territory) with the Ecuadorian state's sovereign rights over the land (as a protected natural area—all of which, according to the constitution, are the patrimony of Ecuador's entire citizenry). Cofán leaders and government officials decided to categorize the RECB as a "coadministered" area, even though the Cofán are solely responsible for protecting the reserve.

The state's rationale for giving the Cofán control of the RECB is relatively straightforward. Ecuador's Ministry of Environment is underfunded, under-staffed, and often ineffective. It is in desperate need of enforcement agents and allies with external funds. The Cofán provide substantial aid in both areas. The immediate Cofán reasons for cooperating with the Ministry, in

Cofán park guards (Photo by Michael Cepek)

turn, are not difficult to identify. Only one of the four Cofán communities in the area had any state recognition before the creation of the reserve.
Absolutely no demarcation of Cofán lands had taken place. Under Ecuador's
Forestry Law, new reserves are created by ministerial decree. In one fell
swoop, the reserve declaration recognized the area as ancestral Cofán territory, and it gave administration and management powers to the Cofán.

At first glance, the arrangement seems rational enough from the Cofán
perspective. In return for the obligation to manage their territory as an
ecologically protected area, Cofán people receive state power and land
rights. Moreover, they receive international financing under the banner that
their territory is not only Cofán land but the ecological patrimony of all
Ecuadorians and the entire world. An additional logic, however, lies behind
Cofán leaders' campaign for coadministration of their ancestral territory.

Even though the FSC and FEINCE are increasingly strong instruments of
the Cofán nation, Cofán solidarity continues to be complicated by the relative
autonomy of households, extended families, and communities. The weakness
of overarching social cohesion, as well as the difficulties of coordination at
the "national" level, explain certain leaders' desires to place their territory
not only under Cofán power but under state power as well. In an unpublished
outline for a national system based on the RECB model, Randy Borman
describes the Cofán vision for state, indigenous, and community relations:

[Coadministered reserves must] establish strict and inflexible "boundaries" for the government, the indigenous group whose territory is involved, and the local communities, concerning any and all strategies that threaten the long-term integrity of the area. . . . These "boundaries" should be accompanied by stiff penalties, the most obvious penalty being the termination of the legal rights of the offending party within the area covered by the *convenio* [agreement]. In the case of the government, this would mean the immediate ceding of absolute long-term control of the ancestral territory to the indigenous group. . . . In the case of the indigenous group, this would mean the immediate ceding of all claims of ancestral rights over the area to the government. In the case of the local community, this would mean the loss of all use and possession rights for the individuals within that community, and entail the moving of the entire community from within the established boundaries.

The legal and political mechanisms for solidifying Borman's vision do not yet exist. The Cofán manage the RECB according to a set of makeshift powers provided under a loosely organized Acta de Acuerdos, which FEINCE signed with the Ministry of Environment in 2003. Even though Borman's statement does not correspond to current political reality, its logic is telling, and it might trouble analysts of indigenous and environmentalist movements. Most anthropologists rightfully view the revocation of ancestral rights and the forced removal of local communities as colonialist forms of conservation. In the Cofán case, however, the provisions emerge not from state or conservationist positions but from the indigenous position.

In effect, what Cofán leaders seek to legislate with the coadministration proposal is an essentialized foundation of Cofán values and Cofán identity. After the creation of the RECB, they established similar agreements for the majority of titled Cofán lands, which overlap with the Cuyabeno Wildlife Reserve, the Cayambe-Coca Ecological Reserve, and the Río Cofanes Territory. According to their logic, the continuity of the Amazonian environment is the ultimate basis for anything or anyone that could be called "Cofán"—past, present, or future. The protection of their traditional territory therefore trumps individual, community, and even national sovereignty. The ultimate goal is to protect the forest against all predators, even if they are Cofán themselves, and even if environmental protection results in an irrevocable constraint on Cofán self-determination.

Of course, it is possible to interpret the Cofán proposal as the product of political and economic desperation. Indeed, contemporary circumstances force Cofán people to search for whatever scraps of power they can find. After years of research and conversation with Cofán leaders, however, I see much more behind their coadministration proposal than political opportunism. From my perspective, the creation of the RECB and similar protected areas reflects an epochal moment in Cofán self-understanding.

In Cofán demands for coadministration of indigenous ecological reserves, the Cofán nation does not appear as an autonomous or fully embodied self. Rather, it exists in two sets of substantive relations: first, between Cofán people and the environment they consider to be essential to their mode of being; and second, between Cofán people and the actors who support their efforts to maintain that environment as well as their relationship to it. Consequently, Cofán environmental politics are not based on an understanding of autonomy as the completely free action of an a priori individual or collective will. Rather, Cofán activists direct their efforts toward the establishment of mutually binding relationships with actors who can help them maintain their increasingly threatened forests.

In effect, the collective subject of Cofán environmental politics is based on a double contract. Individuals and communities cede autonomy to the Cofán nation in return for effective territorial protection. The Cofán nation, in turn, commits itself to an essentialized foundation of Cofán identity in return for legal recognition, enforcement powers, and global funds. Cofán leaders understand such aid as essential to the survival of their rain forest territory, which they want to protect in a way that satisfies their own values as well as the agendas of the Ecuadorian state and international civil society. The Cofán vision of environmental politics is innovative and powerful. It explains much of the FSC's success in attracting international aid and state support for its initiatives.

ACTING AS AN NGO

Organizing as an NGO opened new possibilities for political action. FEINCE and its predecessors provided a unified voice for the Cofán nation. Nevertheless, the federation was never capable of securing the resources that became essential to Cofán politics in the new millennium. Currently, Cofán political action demands constant meetings with state officials, corporate officers, and NGO representatives in Quito. Land acquisition involves large-scale, GPS-supported boundary-marking trips throughout Cofán territory. The educational development of future leaders requires a residential base, as well as financial and logistical capabilities, in Quito. And maintaining constant communication with donor institutions is essential. All these practices are key elements of Cofán people's struggle to become a legally empowered, academically credited, and financially remunerated force for global conservation.

Organizing as an NGO, however, entailed significant transformations of Cofán life. My observations of the changes fall into three areas: Borman's experiences of the social, material, and psychological shifts required by work as the FSC's Executive Director; Aguinda's negotiation of urban bureaucracy and ethnic difference; and the more general demands of accountability and transparency that face all FSC employees. The NGO form is not simply a means to promote preexisting political ends. Rather, its adoption requires significant transformations of the people whose voice and vision it represents.

Institutionalizing Political Action: Property, Practice, and Psychology

Borman's decision to engage in full-time political work in Quito disrupted his contact with residents of Zábalo. Even in Quito, however, Borman is almost always in the company of Quenamá. Moreover, A'ingae continues to be the language of everyday life at the QCC. Nevertheless, Borman worries that his distance from Cofán communities threatens his relation with his social base, which he considers to be his most valuable asset as an indigenous leader:

> One of the big reasons why Quito is so frustrating is because it leaves me out of the social context from which I gain my edge. This is one of my biggest criticisms of the quote-unquote "indigenous leaders." They're no longer part of the social context. And therefore their ability to translate accurately what their constituency is really feeling is lost. You begin to develop your opinions and your ideas in a vacuum. They are no longer translations of the community's basic interests. My major strength is my ability to coalesce a whole attitude and movement. I obviously influence it, but I am also very much a product of it at the same time. I don't want to shut that off.

The FSC's growth has diminished everyday relations between Borman and the people of Zábalo. At the same time, it has deepened the ties between Borman and the greater Cofán nation. In addition, it has created a new predicament: separating "the *fundación*" from "Randy Borman." Outsiders must see the FSC as more than a vehicle for one man's ideas and projects, especially if the individual is as atypical a representative as Borman. As a personal, practical, and psychological issue, however, the challenge requires more explanation.

In 2001 and 2002, Borman and Quenamá lived with their children in the same building that housed the FSC's main office, where employees worked every day.[4] Often, the office door was ajar. FSC staff used Borman's home to relax, to eat, and to meet with Cofán and non-Cofán visitors. Children and guest Cofán constantly moved in and out of the office. They socialized with their parents, aunts, uncles, and classificatory siblings, who, thanks to a flexible kinship terminology, were none other than Borman, Quenamá, and Aguinda. The only location that was relatively off-limits to FSC employees and visitors was Borman and Quenamá's upstairs bedroom. Nevertheless, it was not uncommon to see Lopez and Espinosa speaking to Borman as he sat next to his bed, tapping away on yet another computer. The division between FSC space and residential space was poorly defined, to put it mildly.

When the FSC was created, the blurry boundaries were a financial necessity. Harlow's donated funds financed property acquisition, construction, and remodeling. They did not, however, purchase the furniture and equipment that made the QCC into a place where Cofán students could stay and FSC employees could meet with visiting officials, eat their lunches, and converse with each other around a large table. Most of the objects came from Borman's old Quito residence, which he bought with nearly fifteen years' worth of tourist dollars. Borman estimates that the monetary value of the objects that he purchased and made available to the QCC totals nearly $10,000. In addition, other "private" goods became essential to FSC activities. For example, Borman uses his own truck to attend meetings in Quito as well as to make trips to the lowlands.

When tourism was at its peak, the overlap of individual/family and national/political spaces, objects, and expenses was never a problem. Borman had a steady income and a subsistence base in Zábalo. With a reduced paycheck and an almost entirely commodity-enabled existence in Quito, however, it is no longer feasible for Borman and Quenamá to make their property and income available for FSC activities. Nevertheless, in 2001 and 2002, neither the Borman family nor the FSC had enough money to exist apart from each other. The sum of Borman and Quenamá's salaries was not enough to finance housing, food, and education for their family. And FSC funds could not support a fully equipped workspace or many of the everyday expenses that NGO work requires.

The impossibility of separating the QCC from the Borman household produced many problems at both ends of the relation. Cofán people find it hard to distinguish between the FSC, the QCC, and the Borman family. They sometimes voice suspicions that Borman and Quenamá are using money donated "to the Cofán" for personal gain. The gossip infuriates Quenamá, who often complains that much of her energy and resources go to the FSC. Moreover, all FSC employees realize that the use of the QCC's objects leads to their wear and tear, in effect causing the consumption of Borman and Quenamá's property for Cofán-wide ends.

The impossibility of dividing Borman's personal life from his work with the FSC involves much more than space and property. Borman's ideal existence is set in the forest, where he can practice his twin loves of natural science and hunting while residing with other Cofán people. Although working as the FSC's director has allowed him to maintain the freedom of a non-nine-to-five lifestyle, his time in Quito creates a great political and existential irony: the bulk of his life is consumed by working to protect something that he can no longer enjoy.

In Quito, Borman spends the majority of his life indoors, either in front of a computer or at meetings with government and NGO officials. Increasingly, he suffers from the fatigue and dissatisfaction of sedentary, urban life. He can feel his body atrophying and his forest skills diminishing. Quito's cold, dry, high-altitude environment exacerbates the difficulties. Few Cofán people can withstand life in the capital for more than a month at a time. (After nearly three years of fieldwork with the Cofán, I can state unequivocally that the sentiment is universal.) Although Borman realizes that his ability to conceptualize, fund, and implement projects is an important skill, he resents his increasing isolation from Cofán communities and the Amazonian environment.

In 2001 and 2002, Borman lamented the fact that his life had come to resemble that of a "dorm parent." He woke up early each morning to cook breakfast for the Cofán schoolchildren, and he spent many of his evenings and weekends helping the students with homework. In general, Borman feels that his work as FSC director is all-consuming in terms of the time, energy, and focus it demands. He finds it especially frustrating to be an object of suspicion by Cofán insiders and non-Cofán outsiders, who imagine that he desires NGO work because of its steady income. He has repeatedly offered his position to any Cofán who will take it, and there are never any takers. (Very few Cofán people understand what such work entails, of course, and they know that they do not possess Borman's abilities.)

In 2001, Borman told me that his ideal life as FSC director would be to rotate work and residence on a three-month schedule: one month in Zábalo, one month in other Cofán communities, and one month in Quito. When I suggested that such a life would be the exact opposite of *opatssi* conviviality, he replied that it would be no more difficult than his life is now. At the very least, he suggested, the situation would allow him to spend more time in the forests whose protection is the FSC's primary objective.

In addition to the social and geographical difficulties of FSC work, obeying the structures of NGO life requires a difficult shift in psychological orientations. The FSC has no endowment. It cannot operate unless it finds new funding streams by constantly pursuing grants. To date, the FSC has been fairly lucky, although there have been many crises. When I arrived in

2001, Borman, Lopez, and Espinosa were very worried about the future of the organization. That year, Borman went four months without receiving a paycheck. The resources from one grant ended without another arriving to replace them. The situation was nerve racking for all involved. It made it difficult for any FSC employee to muster the confidence to plan for the next month, let alone the next five years.

An NGO-enabled existence differs strikingly from life in Cofán communities. In the lowlands, cash and commodities come and go. The availability of the forest, however, makes it hard to imagine having no place to live and nothing to eat. Unlike forest life, where a basic assumption of material continuity forms the existential background, living through an urban NGO is ruthlessly contingent. Access to outside resources is the uncertain foundation of everything. Through grant proposals and formalized accounting practices, the acquisition of the most basic necessities must be explicitly ordered, formally justified, and contingently planned. Although the situation is less jarring for such individuals as Lopez and Espinosa, who have always lived in a capitalist lifeworld, it is a different story for the Cofán. Cofán people still produce much of their lives with objects that have not passed through the uncertain pathways of the commodity circuit.

Proposing projects as a means of ensuring one's existence requires tailoring one's needs to the shifting agendas of the aid community. Borman laments the lack of "on-the-ground flexibility." Donor institutions have little sense of the socioeconomic realities of rural Ecuador, and most do not fund administrative expenses. Therefore, grant work almost always implies a disingenuous representation of project costs. NGO employees must find a way to pay themselves while producing an identifiable "product"—the purchase of a piece of land, the cutting of a boundary trail, or the construction of a guard station—that justifies a grant. Furthermore, even the most generous grants cover no more than five years. Many last for as little as a few months. Summarizing the situation, Borman explained:

> I find the basic tenet of *proyectos*—that they end—very frustrating. I never really thought of that as a defining element, but it certainly is. That makes all the sense in the world. Because we have long-term needs, and the nature of *proyectos* is to be short-term. And then to have to go through the whole nervous process of trying to decide whether we're going to have anything continue, all this sort of stuff—it's nerve racking. Personally, I find *proyectos* very, very frustrating as a means of work. Each time it's a different format. Each time it's a different set of demands. I'm probably overly blunt when I talk with project organization people, saying, "It's not what we want—it's

what you guys have to offer." But I've found it usually cuts out three or four days' worth of garbage. I think the whole *proyecto* mentality fosters this grab-it-while-you-can attitude.

Despite the experienced difficulties of working through the FSC, Borman has become a successful NGO director precisely because of his unique skills. Conceptualizing, funding, and operationalizing projects is tiresome and frustrating. Nevertheless, it depends on the creativity that makes Borman an effective political agent. Borman's multicontextual sense of identity is particularly useful for negotiating relations with social others. I have listened to him make English-language appeals to the "common sense" of a North American USAID officer in the presence of Spanish-speaking Quito bureaucrats. I have heard him discuss the pros and cons of different ecological field methods with European scientists. I have seen him hold hands in prayer with members of Quito's overseas evangelical community while speaking about schooling possibilities for Cofán children. And I have watched him engage in long A'ingae discussions with Quenamá, Aguinda, and FEINCE leaders while meeting with oil company spokespeople, who had no choice but to wait patiently as the *indígenas* came to consensus. Borman wears many hats. He can position himself as a sensible American citizen, an educated Western scientist, a committed Christian brother, or an authentic indigenous leader. He is a master at framing his arguments with the linguistic and cultural signs that make the most strategic sense at any one moment, in any one context.

Even though Borman resents spending so much time writing grants and attending meetings, he realizes that the FSC's effectiveness depends on his abilities. The next generation of Cofán leaders must match his capacities if the FSC, as well as FEINCE, are to continue their work.

Negotiating Bureaucratic Action and Ethnic Difference

Most Cofán people realize that Borman is their most important leader. None of them, however, believe that one individual should have so much responsibility. Although Borman is the central actor, other Cofán individuals are important political agents. With regard to the FSC, the second most important Cofán activist is Aguinda. He has been with the organization since the beginning. Aguinda is relatively well educated and fairly proficient in Spanish. He completed elementary school, and he spent a year in Ecuador's army, where he improved his language skills and became familiar with the ways of Cocama. The people of Zábalo elected him vice president many times. He was president for a year in 2000. In addition, he has occupied multiple elected posts with FEINCE.

By the late 1990s, Aguinda, too, began to shift his residence. Near the end of the decade, his brother died in Zábalo. He opted for the traditional Cofán custom of abandoning the village, "so that one does not become sad." The period coincided with the beginning of FSC mobilization. Aguinda took the opportunity to make the community of Doreno his home. He grew up in Doreno, his mother lives there, and the village is much closer to Lago Agrio and Quito, the two sites of much of his work for the FSC.

An examination of Aguinda's work and struggles provides a more complete understanding of contemporary Cofán challenges. He is much more representative of general Cofán capacities than Borman. He speaks no English, he has never lived in Quito for extended periods of time, and he has no higher education. Nevertheless, he works in urban NGO politics, and he knows more about Borman's position than does any other Cofán. Furthermore, ethnic others recognize him as an unambiguously indigenous person.

Whereas Borman can mobilize his gringo identity in relations with Ecuadorian, North American, and European others, Aguinda is left with no choice but to act according to his reception as an indigenous person. Although his unambiguous identity has political benefits, his Cofán-ness complicates his work in important ways. Two of his challenges are particularly telling: the difficulties of working in an urban, bureaucratic space; and the tension and misunderstanding that color his interactions with Espinosa and Lopez, the two FSC employees with whom he cooperates most closely.

Even though Aguinda suggests that his title of "Technical Director" makes little sense, he is very clear about his responsibilities. Much of his work occurs in lowland communities, and one of his primary tasks is to organize the preliminary stages of land demarcation. The work requires plotting out community and reserve boundaries with a GPS unit and accompanying government officials to speak with neighboring peoples and invading colonists. Aguinda can operate in areas that are too dangerous for Borman or any other gringo-appearing individual, as Colombian armed groups control important river routes as well as much of the land abutting Cofán territory. In addition, Aguinda serves as the main logistical coordinator for such supporting institutions as the Field Museum, which directs conservation projects in Cofán communities.

Aguinda works frequently in Lago Agrio and Quito. In the lowlands, he coordinates efforts with FEINCE and negotiates with provincial officials. His activities in Quito, however, typically involve Espinosa, who is much more familiar with the national capital. Aguinda describes the bulk of his Quito work as "wandering." During their travels, Aguinda and Espinosa visit tourist agencies to market Cofán-produced Ecocanoas. They spend many days in government offices, where they drop off and pick up documents and meet with bureaucrats. In addition, Espinosa and Aguinda sometimes collaborate on small grant applications. Although Espinosa often works alone, Aguinda stresses that it is important for officials to see that *a'i* also direct FSC activities.

Borman and Aguinda put tremendous stock in Espinosa's familiarity with Quito as a cultural space. They doubt that any Cofán could match the Cocama's success in managing the foreign social world. When I asked Aguinda if he believed that a Cofán person could replace Espinosa, he replied in the negative. He explained that *a'i* cannot live in the capital because of the cold weather, the different food and language, and the absence of forest and community life. He stressed that they would miss "residing with their own *a'i*, telling stories, laughing, doing everything like that." According to Aguinda, a Cofán could not negotiate the daily tasks and unplanned problems that Espinosa constantly manages.

Less than fully aware of his unique abilities, Espinosa believes that the FSC could eventually become an entirely Cofán-managed organization. Unlike Aguinda, he holds that sufficient time and resources would enable him to train Cofán people to take over his work. As with Aguinda, however, Borman doubts the sensibility of Espinosa's position:

> Freddy has incredible energy. He has incredible contacts. He's able to take care of many things that otherwise we would need to hire a lawyer for, and he's just able to handle them. The bottom line is that, at the bureaucratic level, he has connections throughout this city that no one other than a person born into that could ever have. He has the thousand and one little friends in this place and the friends in that place. He understands the mentality of people. It's just this networking ability that no Cofán is ever going to be able to match. No white man. No Indian. No *guayaquileño*.[5] The bottom line is that there's nobody who can match that sort of ability other than a person who's from the same culture. It's dealing with a cultural unit. A whole culture has been developed to handle this urban, Latin, Ecuadorian, *sierra* Quito environment. It's focused. It's focused so tremendously. So we get somebody like Freddy. If I'm going to pick somebody to handle Ocomari Mountain [in western Cofán territory], I give the job to Ciro [a Cofán man from the foothill community of Alto Bermejo]. And if I'm going to pick somebody to handle Quito, it's Freddy. That's good sense.

"Handling Quito" involves an immense amount of implicit social knowledge. Aguinda explained, "*A'i* don't know. Truly, to go everywhere, to go to all of the officelike places, to go and do everything, all of the work that is required—*a'i* don't know." Lopez told me that Cofán people are simply not "animated" enough. In the most literal sense, Espinosa's energy enables his success as the FSC's Director of Communications. He bounds from office to office, documents in hand, and he drives recklessly through Quito traffic. He is a master of politely (and effectively) requesting attention from the secretaries and officials who make FSC activities possible.

As someone born into middle-class Quito society, Espinosa is aware that he embodies a certain style of sociality: a way of meeting, speaking, and relating to other *quiteños*. His skills mean the difference between succeeding or failing to push through a bureaucratic procedure. Even though he believes that he could teach a young Cofán person to act in the same ways, he realizes that getting Cofán to break free from their *opa* moorings— their shyness, their embarrassment, their desire to avoid conflict and tension, their fear of becoming flustered while speaking Spanish—would require intense instruction. When I asked him what such training would accomplish, he replied:

> Well, the fact of knowing how things get done in these offices. Because I'm not just speaking about relations with people. But knowing that after going to this desk you need to go to this other desk, and then to these other desks. And in this other desk, you were talking before with this person. And if you don't see her, you sit there anyway, even though you know that she hasn't arrived. And you do this in order to let people know that someone is behind this document. And when the person does arrive, it's not just one document under the twenty others that are already waiting, but something else. They know that there's someone that is there all the time. And so they say, "Ah, this, yes, I've finally gotten to it." So when I arrive, it's like the whole world knows what's going on, do you understand? And then we go on to the next step. And when they see me, they check and say, "Sir, but this part here is incomplete." Or a signature is missing somewhere. And then I quickly grab it and I leave. And I run to pick up the car and I try to do it as quickly as possible. Not taking the main streets when there's traffic, but taking the short cuts. And then I go on to the next step.

Espinosa and Lopez are committed to the FSC and its mission. Espinosa began working with Borman on Zábalo's tourism operation as a part-time, motorcycle-riding messenger in the capital. Aware of Espinosa's increasing knowledge of Cofán politics and Quito bureaucracy, Borman offered him the FSC position when the organization was founded. Shortly after, Lopez joined the FSC as its accountant. Espinosa and Lopez are evangelical Christians. In addition to trusting Borman as an NGO director and indigenous leader, they view him as a "man of God." Although they have come to feel profound affection for Cofán people, the Cocama continue to consider FSC work as a form of missionary-like *servicio*. They see the Cofán as a noble but disadvantaged social group. Understanding their work as *servicio* justifies the long hours, the low pay, and the uncertain future.

Even though they have positive opinions of the Cofán, Espinosa and Lopez share many of the assumptions of Ecuador's middle-class mestizos. At times, they argue that the Cofán must "advance." Lopez once told me that Cofán people have only recently begun to "evolve" in a psychological sense. From her perspective, they are still very manipulable, very *inocente*. For Lopez, their "innocence" connotes at least partial ignorance of project implementation. She is especially critical of Cofán difficulties with financial management. She believes that their misunderstandings explain Cofán people's suspicions of her work as well as that of Espinosa and Borman. (Lopez is aware that some Cofán people believe that she is taking project money for herself.)[6]

Although they are conscious of the fundamental differences between mestizo life in Quito and Cofán life in the lowlands, Espinosa and Lopez do not realize how difficult it is to explain FSC projects to Cofán people, whether average community members or experienced leaders. For Espinosa and Lopez, many Cofán stumblings, such as forgetting to sign certain forms or failing to turn in receipts, appear to be a basic Cofán "intransigence." Often, they express visible frustration with Cofán difficulties.

In the eyes of Cofán people, such exasperation appears as anger and unfair criticism. Cofán social norms suppress overt critique. There is a tremendous pressure to avoid structuring situations with explicit hierarchies. No matter how well intentioned Lopez's and Espinosa's criticism, the Cofán experience it as insulting. They think that the Cocama relate to them as "fools" and "idiots." Aguinda has learned to deal with their impatience. Nevertheless, the Cofán students who reside at the QCC feel that Lopez and Espinosa constantly accuse them of being lazy. They suggest that the Cocama believe that they cannot think properly. Based on my own observations, the criticisms of Lopez and Espinosa are rarely overt. Cofán people, however, disagree. Their interpretation creates substantial tension between the Cocama and the Cofán at the QCC.

Less explicitly, Lopez's and Espinosa's discourse betrays their class and ethnic background. Espinosa is overly confident in his discussions with Aguinda. He is louder and more direct. He introduces the great majority of topics into conversation, and he typically shifts topics and declines to follow Aguinda's leads. Espinosa assumes his superior competence in areas that Aguinda has overseen for years, such as the organization of trips to Zábalo or logistical planning for the Field Museum. Aguinda, however, submits to the discursive hierarchy. He often sits quietly as Espinosa proposes a plan. Only after listening to the Cocama does Aguinda suggest that there might be a better strategy. He also shows deference to Lopez. During many conversations, he grins silently while Lopez criticizes Cofán work practices.

In the grand scheme of things, this treatment does not bother Aguinda, who has become intimately familiar with Cocama ways. Nevertheless, it distorts his interpretations of Lopez and Espinosa's goodwill. As with other Cofán, Aguinda assumes that Lopez and Espinosa see the FSC as little more than a paycheck. He doubts whether the Cocama actually enjoy the company of *a'i*. For example, he often critiques Lopez (behind her back) for failing to follow protocol in greeting and conversing with Cofán visitors.

In general, there is a lack of confidence at the FSC. The distrust follows ethnic lines. The Cocama feel that the Cofán could be working harder to complete tasks in the required ways. And the Cofán feel that the Cocama dislike them and mainly desire cash. Nevertheless, the FSC continues to function effectively. In large part, Borman provides the cohering force. He is the one individual whom all parties trust to act in the appropriate ways.

Transparency and Accountability

Wavering confidence characterizes relations between FSC employees. A much more significant form of distrust, however, exists between the FSC, donor organizations, and the Ecuadorian government. Over the last decade, the World Bank and the International Monetary Fund began to use the discourse of "transparency and accountability" to diagnose the problems of such unstable nation-states as Ecuador. In 2001, Transparency International named the country as one of the world's ten most corrupt nations. Shortly after the millennium, Ecuadorian officials began to speak about a "new tax culture." They declared their intention to make tax laws apply to citizens and corporations in a more rigorous fashion.

At the time, some Ecuadorians paid no taxes at all. Many businesses worked without formal receipts or legal accounting systems. Calls for transparency and accountability went beyond financial matters to engage the murky nature of the Ecuadorian state. Articulating the critique from the perspective of North American capital, E. Anthony Wayne, the Assistant U.S. Secretary of Business and Economic Affairs, explained: "The problems that U.S. investors in Ecuador have reported to the State Department . . . include 'unclear and contradictory policy signals from competing governmental entities, inconsistent implementation of laws and regulations affecting foreign investors, and opaque judicial processes'" (Wayne 2004).

The FSC receives a significant portion of its funding from USAID, which is attempting to make Ecuador more suitable for stable democratic rule and "robust" economic development. As a recipient of USAID aid, the FSC needs to comply with increasingly strict rules for financial management. Compliance, however, creates problems for Cofán political action. By examining the FSC as a "social space defined by transparency debates" (Marcus and Powell 2003:327), the intersection of global power and daily life becomes apparent. For the Cofán, engaging with a new regime of transparency and accountability entails anxious attendance to a novel set of demands. From their perspective, the requirements can paralyze their work and minimize their accomplishments.

In 2001 and 2002, the majority of FSC funds came from the Field Museum and USAID. As a private U.S. institution, the Field Museum has its own financial requirements. Its standards mesh well with the practicalities of Cofán life and politics. The organizations that administer USAID money, however, work in different ways. During my fieldwork, the main intermediaries between the FSC and USAID were the International Organization of Migration (IOM) and CAIMAN.[7] IOM awarded $60,000 to the FSC in 2001 and 2002. The money arrived initially through Ecuador's state program UDENOR (Unity for the Development of the Northern Border). Its purpose was to fund land demarcation and infrastructure development in Cofán territory. USAID awarded CAIMAN a budget of $6,502,000 for the years 2002 to 2005. The Cofán were to receive an unspecified portion of the funds.

As an institution that intends to strengthen the Ecuadorian state, to protect U.S. investments, and to placate domestic critics of ineffective aid programs, USAID uses the language of transparency and accountability to describe its goals. A USAID data sheet described its Ecuadorian objectives in the following terms:

> [There are] five strategic areas for which USAID is requesting 2004 and 2005 funds: biodiversity conservation; democracy and governance; economic opportunities; and development of Ecuador's northern and southern borders. USAID supports the conservation of biologically important regions within Ecuador's protected area systems. USAID also seeks to increase support for the democratic system by strengthening the transparency and accountability of democratic institutions, fostering greater inclusiveness of disadvantaged groups in democratic processes, and increasing consensus on policies critical to democratic consolidation. USAID aims at reducing rural and urban poverty by helping to develop a strong and sustainable microfinance sector in Ecuador and by improving the macroeconomic environment for more equitable growth. USAID continues working with the Government of Ecuador to contain the spread of a coca/cocaine economy into Ecuador by supporting

the construction of social and productive infrastructure projects and providing alternative income opportunities for small and medium-sized farmers. (USAID 2005)

Before the FSC began receiving USAID funds, its accounting practices were relatively straightforward. Lopez relied on familiarity, face-to-face communication, and trust to disburse cash and record and justify expenses. She gave Aguinda and Borman the money that they requested for specific tasks, and she did not require them to return receipts for their costs. She trusted Borman, and Borman trusted Aguinda. In addition, there was a generalized knowledge of recurrent expenses: bus rides between Quito and Lago Agrio, gasoline for canoe travel between Cofán communities, and the basic money required per day (for food and lodging) for individuals conducting project work in the city and the countryside.

Lopez kept a spreadsheet of the income and expenses. If Borman, Aguinda, Espinosa, and Quenamá confirmed that planned work was completed, there were no problems. Lopez did turn in statements to Ecuador's Internal Revenue Service and Superintendent of Banks, but the forms were relatively simple. They consisted of monthly reports of income and expenses, monthly declarations of retentions from FSC employee paychecks, and a 12 percent value-added tax on the FSC's purchase of "services" from Cofán individuals who worked on project-related activities. Lopez had the option of bringing more Cofán into the FSC as regular employees. Doing so, however, would have required retaining 21 percent of their income for social security, of which they would make little use.

When the FSC began working with the IOM in 2001, accounting conflicts emerged. Lopez turned in expense reports with supporting materials and some receipts. The IOM accountant told her that they would not reimburse many of the costs. The main problem was the FSC's inability to produce an adequate paper trail. The IOM wanted the FSC to submit *facturas* (state-issued invoices) as material records for all purchases of goods and services over $5. In order for an individual to be able to offer a *factura*, they must have a *registro único de contribuyente* (RUC; unique registry of taxpayer). A RUC is an official number identifying a natural or corporate person for purposes of tax collecting. In 2002, having a RUC enabled a person to buy one hundred *facturas* from the Internal Revenue Service for $20.

In 2001 and 2002, many Ecuadorian businesses did not possess *facturas*. Ecuador has a large informal and semiformal economy. Many businesses that regularly give receipts, not to mention those that give none, are not *facturable* (working through *facturas*). *Facturas* provide a reliable means of state surveillance for all economic activity. Informal receipts can take the form of a scribbled piece of paper or a printout from a cash register. Their existence, however, does not imply direct state intervention. Many businesses do not want to sacrifice customers who need *facturas*. Accordingly, they obtain illegally printed counterfeit *facturas*.

The main reason businesses hesitate to use *facturas* is that any transactions made with them must include a 12 percent value-added tax. Customers do not want to pay the tax. Businesses do not want to lose customers who can seek another merchant that charges less because it does not issue *facturas*, and thus does not pay taxes. Some businesses will provide *facturas* only if the customer pays a surcharge above the advertised price. Moreover, many Ecuadorians are hesitant to sign *facturas* or any other papers whose destination is uncertain. Functional illiteracy continues to be common in the country, and many people are suspicious of anyone who demands their signature. The problems are especially severe in such provincial cities as Lago Agrio. They are even more common in small towns, not to mention indigenous communities. In some cases, a business would rather refuse a sale than offer a *factura* or receipt of any kind.

By not reimbursing Cofán project workers who do not possess *facturas*, USAID hopes to discipline them. It wants to make the entirety of Cofán actions and transactions visible to itself and the Ecuadorian state, which can thereby increase its tax revenues. From the Cofán perspective, the demand makes them responsible for transforming an entire socioeconomic field. The majority of Cofán people who work on FSC projects find it difficult to engage merchants with the requisite intensity. Cocama dismissiveness of *indios* who barely speak Spanish exacerbates the problems. In Lago Agrio, Cofán people experience many difficulties with *facturas*. They often have to argue with business owners who do not want to give the receipts. Sometimes, merchants charge more if they do supply *facturas*. Cofán people cannot reliably assess the legality of the *facturas* that they receive. And they cannot adequately explain the legal and political rationale behind their demands in order to convince merchants to give them *facturas*.

Although the complaints might seem petty, they have a significant impact on project work. When the FSC is coordinating a boundary-marking trip in Cofán territory, it calls on individuals from all communities to participate. It offers to pay for transport, food, and lodging. In order to congregate at a central point, individuals need to take bus and canoe rides. Moreover, they must eat and sleep along the way. Each step requires a commodity transaction, and each transaction must be supported by a *factura* if an individual is to recoup their costs.

In Sucumbíos, requesting *facturas* is difficult. When Ecuador's Internal Revenue Service cracked down on tax evasion in Lago Agrio in 2003, it closed nearly every business except for the large oil- and tourist-oriented hotels (Lopez, personal communication, 2003). For Cofán who travel from communities located along the San Miguel River, the difficulties are especially severe. Often, the best place to buy food and materials is on the Colombian side of the border. In 2001 and 2002, the area had little government presence due to FARC control. Getting receipts for transactions in the region was nearly impossible. Moreover, making suspicious requests of Colombian Cocama, whom Cofán believe to be "killers," was especially stressful.

Requesting receipts for all transactions implies a basic shift in Cofán habits. Even Borman and Aguinda claim to have problems requesting, inspecting, and holding on to *facturas*. When Aguinda is in the lowlands coordinating a boundary-marking trip, he pays for the transportation and meals of as many as fifty Cofán. He finds it very difficult to collect and keep dozens of receipts. Paper of any kind has a short-lived existence in Amazonian Ecuador, especially when it is kept in the pocket or backpack of someone who is swinging a machete and hiking through rain-drenched forests for a month at a time. Moreover, even though the IOM demands *facturas* for all transactions over $5, there are many goods and services that cost less but that add up over time. The IOM will reimburse small expenses if accompanied by informal receipts. Nevertheless, as Borman explained, obtaining small receipts is a trying experience:

> You know, you need $10, you reach into your pocket and you pull out $10. Neither I nor Freddy is very good at keeping detailed records for a few cents. And we wind up using substantial amounts of our salaries for *fundación*-oriented activities. I was so used to having enough money in my pocket from tourism that I was able to handle all these expenses without having to keep receipts. It wasn't a big deal to give people a lunch in Quito. I definitely miss that. A state where we were doing well enough that we didn't have to keep receipts for pennies because some organization was going to come down on us. [What's so difficult about it?—M. Cepek] It's just time-consuming and a bother. I'm in a section of the city where I definitely need to have somebody guarding the car, so I give a guy on the street $.25. I get in my car, and to stop and write down that $.25 while I'm trying to pull into traffic—it just doesn't occur to me. But by the end of a month of parking, you've probably parked four times a day, and you've been out on the roads fifteen times—it's sixty times that you've paid somebody $.25. When you're getting fairly low salaries to begin with, that begins to hurt. When you're

talking a couple of thousands dollars a month it's no big deal to be using $15 for parking tips. But when you're down to $400 or $500, that suddenly becomes a fairly significant portion.

When intermediary organizations detect a problem in FSC accounting, they halt reimbursement and delay the advance of new funds. They can call for an investigation and put project work on hold. For groups such as the Cofán, the difficulty can be severe. If fifty Cofán people from thirteen communities converge to cut boundaries around an upriver territory, they cannot simply return to their villages to wait for more funds. Their departure would be a serious logistical and financial challenge, given difficulties with communication and transportation.

Considering the issue, Lopez reflected on the motivations of the IOM and CAIMAN: "They aren't here to help, Mike. Really, they're here to paralyze." FSC employees question the true mission of grant-implementing organizations. They assert that the institutions create a "need" for themselves by portraying local NGOs as untrustworthy and incapable of unsupervised action. If more money went directly to such organizations as the FSC, large domestic and international NGOs would no longer be necessary, and their employees would lose their jobs.

The disruption of project workflow is the main reason why FSC employees describe their foremost grant-related desire as "flexibility." With administrative costs remaining constant and project costs demanding sustained action, the FSC cannot put its operations on hold while an intermediary investigates its finances. Often, the only option is to take funds that came in for a different project and use them for the stalled work. FSC staff refer to this practice as the "lending" of funds.

Unplanned costs sometimes emerge with no warning. Long after the completion of the QCC, the builders called the FSC and requested more than $10,000 in initially unsolicited value-added taxes. Under the threat of legal action, the FSC had to find the money in other accounts on pain of shutting down its operations. Some funding agencies are open to "lending." Their main demand is the completion of the projects for which they originally awarded grants. In other words, they do not desire to witness and discipline every step of project implementation. The Field Museum, for example, does not examine the legal details of the FSC's paper trail. Instead, it requires the FSC to provide regular reports of income and expenses. Moreover, Field Museum staff regularly travel to Cofán territory to view the products and processes of project work. The IOM and CAIMAN, however, work differently. The organizations pursue the larger USAID objective of making Ecuadorian society suitable for surveillance and tax collection.

In 2002, problems with *facturas* and flexibility created a conflict between the FSC and the IOM. The IOM first disbursed $23,000 to the FSC for building guard stations and making boundary trails. Ultimately, the FSC used the money to construct two twenty-kilometer trails, to do all of the GPS mapping in Zábalo and two other communities, and to build three park guard stations near the community of Sinangoe. But the IOM refused to hand over $37,000 of the original grant. It cited a number of problems: very few of the FSC's receipts were *facturas*; a handful of receipts were altered, such as one in which a group of Cofán changed a hotel bill from $8 to $18; and the FSC temporarily "borrowed" a portion of the initial funds to purchase a truck. (The FSC's old truck [i.e., Borman's old truck] broke down, and the foundation was left without transportation in Quito.)

The FSC responded to each of the allegations not by denying their validity but by explaining the logic of its actions. It justified the great bulk of *factura*-less expenses with simple receipts for Cofán construction of the guard stations. It paid community members $9,000 for all of the work. The amount was miniscule compared to the bid of a *facturable* Quito construction firm, which put the cost for materials alone at $25,500. The altered receipts accounted for a very small percentage of project costs. A group of largely illiterate and incomeless upriver Cofán changed the $8 receipt because they were able to obtain a receipt for lodging but not for transportation, so they combined the two amounts on one receipt. And the FSC quickly paid back the money they "borrowed" for the truck. Without the vehicle, they would have completed little if any of the work.

At the end of the IOM-sponsored work, FSC employees decided to minimize CAIMAN problems by proposing two changes. First, they requested that CAIMAN hire an FSC accountant to deal specifically with CAIMAN-supported projects. (CAIMAN did make the hire.) And second, the FSC selected a member of each Cofán community to receive an official RUC as a contractor. Having a RUC allowed the individuals to offer *facturas* to the FSC and to pay other community members for "liquidations of services," which do not require *facturas*. In effect, the move created a new social standing for Cofán individuals and communities.

The FSC was able to patronize each Cofán contractor as a tractable business. The "business," in turn, could hire community members on a day-by-day basis rather than as regular employees. The problem for any individual who has a RUC and *facturas*, however, is that they must make monthly declarations of value-added tax at a bank in Lago Agrio. If they do not comply, back taxes will accrue interest, and the individual will fall into increasing debt. From the perspective of community contractors, the structure is very confusing. Moreover, it is extremely difficult for any of them to make regular trips to Lago Agrio.[8] In effect, the creation of *facturable* Cofán contractors is a stopgap measure. It has shifted the burden of visibility from the FSC to individual community members.

From the perspective of the FSC, the new regime of transparency and accountability appears to be about nothing more than paper. In the words of Lopez, USAID and its intermediary organizations "don't want to see products—they just want to see papers." Long-term visions and actual accomplishments appear unimportant relative to the primary goal of producing a standard, legalized paper trail. It does not matter if the numbers are too high, if the excess of a raised price goes as a kickback to a friend or family member who owns a *facturable* business, or if actual products are never seen or produced. Regularizing surveillance and state revenue is the major mission.

At a meeting with USAID officials about the IOM problems, Borman appealed to "common sense." He argued that no one should worry about the details of a $300 project that involved thirty people cutting trails over ten days. And no one should decry the absence of *facturas* when Cofán community members constructed a backcountry guard station for $3,000, especially when a Quito firm estimated the costs for materials alone at $8,500. Lopez suggested that there are alternatives to paper trails. "They have one way of seeing," she stated, "and I have another—seeing things themselves." Participants can write reports, take photos, or produce a video of actual products when *facturas* are unavailable. If the additional measures are insufficient, Lopez explained, "environmental audits" are possible. She suggested that "the people, the administrators, the accountants, and the financers take off their shoes and put on their boots and enter the countryside, making a balance between those who finance and what they finance."

Lopez's ideas point to another way of bridging the distance between the watching eyes of donors, the recipients of their support, and stated project objectives. The alternate way of "seeing" avoids the primacy of state-issued financial records, whose absence can paralyze project work while making other forms of corruption possible. Lopez's ideas are built on a less mediated and less manipulable form of encounter. Evaluators would judge projects not by paper traces but by their social and material products. Ultimately, FSC staff believe that they should not have to sacrifice their mission in order to make the entirety of Ecuadorian society transparent and accountable.

PERSPECTIVES FROM ZÁBALO

Financial crises, urban isolation, ethnic tension, and accounting obstacles are daily concerns for Borman, Quenamá, and Aguinda. The difficulties of living through an NGO, however, are beyond the experience of most Cofán people. As I explained above, Cofán consider their *na'su* to be ethnically ambiguous, morally ambivalent, and problematically oriented to external contexts. They extend their understanding of shamans and contemporary *na'su* to FSC employees. The distance and novelty of the organization make its workings opaque. Zábalo residents commonly declare their ignorance of the FSC. Few claim to know what it is, how it began, or what it is doing. When pressed, however, many of them do have something to say about the *fundación*.

Borman and Aguinda discuss many FSC matters in community meetings. Cofán people know that Borman is the *fundación na'su*. They also know that the FSC gives paid work to Aguinda, Quenamá, Espinosa, and Lopez. Many realize that money from "other lands" supports FSC salaries, buildings, and vehicles. Most speculations concerning the FSC revolve around cash. The most suspicious individuals believe that the FSC is using them. They know that without claiming to support the Cofán, FSC employees would not be living a commodity-based urban life.

Zábalo residents know that most of the FSC's initiatives involve *ande* (land). Borman has repeatedly told community members that the FSC "is fighting for land." In addition to listening to Borman's explanations, Cofán people learn about the FSC's role in territorial acquisition by participating in boundary-marking trips. Furthermore, they know that the *fundación* supports Cofán education in Quito. Cofán people conceptualize the FSC as something that "helps" by providing institutional sites—*proyectos*—for the transfer of work, goods, and services to communities. The more long-term projects, such as the Field Museum's biological monitoring program in Zábalo, take on a relatively autonomous existence. One Zábalo resident differentiated a *proyecto* from a *fundación* by saying that the former is "a thing that ends." The latter, in contrast, is "a thing that does not end." In other words, community members differentiate projects from *fundaciones* by their durability. As the main example of the latter, Cofán now understand the FSC as a relatively permanent fixture of their political life.

Cofán people assume that FSC employees operate by mastering an alien domain of knowledge and practice. Most Cofán focus on one aspect of the required knowledge, and they lose sight of the more complex processes through which the *fundación* accomplishes its work. For example, because most Cofán feel that proficiency in Spanish is beyond their reach, they believe that linguistic mastery alone is sufficient for negotiating relations with outsiders. When thinking specifically of the FSC, writing becomes the paramount focus. Some Zábalo residents believe that they themselves could do the work of Borman, Espinosa, and Aguinda if they knew the language of pen, paper, and computer. (Their fetishization of writing, incidentally, is similar to USAID's own obsession with paper.)

Community members assert that FSC employees engage in the basic process of *fuite'chove iñajañe* (requesting help) by *ttevavaeñe* (writing) their *in'jan'cho* (desires) and *injienge'cho* (necessities), which they then *moeñe* (send) or *jacamba afeye* (travel and give) to the government and other *fundaciones*. The outside actors then offer the *fuite'cho* (help) to the FSC, for which the FSC must *firmaye* (sign). Cofán gloss the whole sequence of actions as *na'sundeccuma noña'ngiañe* (bothering leaders). If completed correctly, it terminates in the *isuye* (picking up) or *panzaye* (amassing large amounts) of *fuite'cho* (help: basically, money, material goods, jobs).

Cofán people view the acquisition of resources as the main reason to interact with outside NGOs and the Ecuadorian government. The disciplinary arm of the state is difficult to discern in Zábalo. People understand the government as a source of goods and services rather than a center of control. Through such federal institutions as the Ministry of Environment, Cofán communities receive land titles. Other arms of the state bring health care, school materials, and outboard motors to Zábalo. NGOs also supply aid. For example, the Field Museum provided income and training to individuals in Zábalo for the better part of a decade.

I interviewed a large portion of Zábalo's residents about *fuite'cho*. Unlike Borman, most of them do think of themselves as objects of aid. Many, however, have no idea why outsiders want to help. Community members suggest that both Gringo and Cocama know that they are "poor and uneducated." People speculate that because some outsiders are "good people," they want to help by sharing their wealth. Some of the more cosmopolitan and politically aware residents focus on the scientists and the ecotourists. They assert that outsiders want to help Zábalo because the community has a reputation for protecting its forest. One man said, "They come here because we have a lot of forest, as it was before. Here we have all kinds of trees and natural medicines, as it was before. That's why." In other words, community conservation practices[9] make Zábalo unique. They attract the sympathy, respect, and admiration of foreigners who want to help the community to accomplish its goals.

After five hundred years of colonialism, political marginalization, and socioeconomic inequality, however, most Cofán cannot imagine that selfless altruism is the ultimate motivation of any outsider. They often wonder why ethnic others come to their village, whether they offer aid or not. They ask themselves: Could the outsiders be selling pictures or writing books about the Cofán, making millions of dollars? Are they students who come to the village, stay for a while, write their thesis, get their degree, and then find good jobs in their home countries? And aren't they all employed in some way, and thus "helping" because someone is paying them?

The assumption of exploitation also influences Cofán opinions of Borman, Quenamá, and Aguinda. No matter how clearly FSC staff explain the difficulty of their work and the level of their expenses, community members find it impossible to imagine that the situation is fair. The interpretation leads many Cofán to "insult" the FSC *na'sundeccu*, just as they have spoken badly of all leaders before them. Nevertheless, their eagerness to criticize FSC employees does not mean that community members do not appreciate the organization or that they would trade their own lives for employment with the *fundación*.

Borman once told me that "thick skin" is a necessary quality for all successful Cofán leaders. Both he and Aguinda have become relatively good at ignoring community bad talk. For the moment, their desire to do political work outweighs the pain of being an object of gossip and insult. Quenamá, however, told me many times that she is at the breaking point. In a field note from 2002, I summarized one of our conversations. Quenamá was discussing the difficulties of working for the FSC as well as her preferred path for Felipe, her oldest son:

> This afternoon in the office, I had an interesting conversation with Amelia. We were talking about the foundation and the difficulties involved in the work of Randy and Roberto. The problems mainly revolve around the fact that people from home communities are constantly *áfase'je*, or "saying bad

Amelia Quenamá (left) with other Zábalo residents (Photo by Michael Cepek)

things about" them. Amelia said that people never appreciate their work or their help. Instead, they assume that something dishonest or manipulative is going on. It's clear that Amelia resents the situation. She thinks that she and Randy and the other foundation workers are being wronged by other Cofán. She said that she told Felipe that he should get a good job when he grows up and work for himself and his family rather than for the Cofán or Zábalo. Her feeling was in reaction to what Randy has gone through. She said that Randy, however, has always maintained his commitment to *fuiteye* or "help" the people as a whole, which she understands but sometimes disagrees with. She also talked about all of her and Randy's money and resources—all that is *ingimbe* or "ours"—and the fact that they've used it for the foundation and the good of people who aren't members of her family. She doesn't think that this has been recognized or appreciated by other Cofán.

Amelia is very convinced of the inevitability of gossip and insult. She said that it's *a'i canse'cho* (Cofán custom), and she can't deal with it any longer. So, Amelia wants to live in her own home, with her own children, away from the QCC, which she thinks should be taken care of by a different couple, Cofán or otherwise.

In our conversation, Quenamá described the especially difficult situation of FSC employees, who suffer a double absence of *opatssi* life. Not only must *fundación* workers accept the burdensome demands that NGO practice places on their everyday lives, they must also suffer the criticisms of a community that no longer accepts them as equals. Creating a more generalized awareness of the political necessity of the FSC is an important challenge. Otherwise, community criticism will make individuals such as Quenamá give up on the *fundación* altogether.

Despite their skepticism and complaints, the majority of Zábalo's residents are glad that the FSC exists. They realize that a secure land base is essential for their future. Moreover, they appreciate the goods, services, and work that the *fundación* brings to their community. Finally, they understand that the formal education of Cofán children is a necessary project. The FSC is the only agent that can secure access to the resources on which Cofán politics depends. Nevertheless, the NGO continues to be an object of mystery. Its association with distance, difference, and power calls forth old suspicions. Cofán people's instinctive distrust forms the interpretive horizon within which they struggle to come to terms with the necessity of organized action.

For the Cofán individuals who work for the FSC, the NGO form is more than an object of speculative critique. It is a structure of practice that is notable for its political potentials and related pains. Cofán activists have learned that NGO work is a necessary evil. The FSC represents a mode of collective action that can make *opatssi* conviviality possible in the contemporary global order. Simultaneously, it demands participation in a form of life whose social, practical, and psychological consequences make Cofán mobilization a politically productive but inherently troubling endeavor.

COLLABORATING WITH SCIENCE AND CONSERVATION

Although many Cofán people benefit from FSC-supported initiatives, few play a direct role in running the organization. At the community level, however, a great number of Cofán are active participants in experimental environmental projects. Zábalo is a case in point. Since the 1980s, its residents have worked to make their village an innovative force for conservation. They now direct one of Latin America's most successful community-based management systems. Publicity regarding their success has attracted a series of NGOs to the village. Most outsiders want to learn from, support, replicate, and improve Cofán practices of *tsampima coiraye* (caring for the forest). Although residents desire external help, their objectives and experiences diverge from the understandings of Western conservationists. More often than not, outsiders come to Zábalo with assumptions that do not match Cofán aspirations.

In this chapter, I explore Cofán conservationism as an everyday practice and an overarching vision. Cofán people relate to the Amazonian environment in an intricate manner. Their environmental practices, in turn, mediate their modes of relating to each other as individuals, as households, as communities, and as a nation. As they learned to value life in an intact environment by witnessing large-scale ecological destruction and involving themselves with ecotourism and scientific research, they institutionalized new forms of protecting their forest, which entailed novel modes of relating to each other and to outside actors and institutions.

Of all the allies that the Zábalo Cofán have attracted with their system of organized forest care, the most important has been the Field Museum of Natural History. Museum personnel worked to document, publicize, and improve Zábalo's management system by applying scientific logic, methods, and instruments to Cofán environmental relations. Rather than embracing the museum's program in toto, however, the Cofán made the institution's training and resources an essential part of their own vision for a just and effective form of indigenous-environmentalist collaboration.

The word and concept of *environment* has no clear analogue for the Cofán. Their closest term for such Western notions as "environment," "wilderness," or "nature" is *tsampi*. Depending on context, *tsampi* has multiple meanings: encompassing, uncleared land (as opposed to *canqque* [community]); standing or regrown forest (as opposed to *nasipa* [horticultural field]); and terrestrial forestland (as opposed to *na'e* [river environment]). The underlying sense is "land with forest cover." Qualifying a space as *tsampi* indicates the absence of large-scale human conversion. In addition, *tsampi* includes all the plant and animal species that make the forest their home. Cofán people contrast *tsampi* with urban areas, where a nonindigenous way of life predominates. Accordingly, they call themselves *tsampini can'jen'sundeccu* (dwellers of the *tsampi*).

Cofán people depend on a limited number of commodities. Basic necessities include salt, soap, and machetes. Some people extend the list to include clothing, pots, axes, matches, guns, ammunition, fishhooks, fishing line, Western medicines, chain saws, and outboard motors (powered, of course, by gasoline). Nevertheless, every person in Zábalo obtains the great majority of their food from grown, collected, hunted, and fished sources. All domestic groups live in dwellings constructed mainly from locally obtained trees, vines, and leaves. And every household produces many essential items (e.g., canoes, banana drink stirrers, and toys) from available forest resources.

A Cofán house (Photo by Michael Cepek)

When income dwindles, Cofán people are prepared to do without. During long conversations about money and its unstable presence in Zábalo, many residents told me that they are capable of returning to the ways of poverty: making salt and soap from vegetable sources; fashioning clothes from bark cloth; catching fish with nets and natural poisons; paddling and poling canoes; and relying on spears, dogs, traps, and blowguns for hunting. Historically, isolation from trade networks was always a possibility. If circumstances demand, Cofán people can produce nearly all they need.

The diversity of species that Cofán people use is impressive. In Doreno, the ethnobiologist Carlos Eduardo Cerón confirmed earlier research (Pinkley 1973) by cataloguing the use of nearly 300 plant species (Cerón 1995). Borman estimates that the Cofán utilize approximately 250 plant species for medicinal purposes and eat approximately fifty species of terrestrial mammals, fifty species of fish, and hundreds of species of birds (2002). Although less impressive in its diversity of cultigens, gardening knowledge is also extremely important. Plantains and bananas supply the overwhelming majority of carbohydrates. In addition, manioc, fruit, and corn are staples. A functioning household is never without certain products, such as banana drink, and it is extremely uncommon for a family to go two days without eating meat or fish.

The importance of the *tsampi* extends beyond its provision of material goods. Many individuals receive "play names" because of physical or behavioral peculiarities that are similar to animal characteristics. Two of my closest Zábalo friends are "Hummingbird" and "Armadillo," and my ritual son is "Leaf-cutter Ant." Children's games revolve around themes of hunting and animal habits. Cofán cosmology describes the actions of "everyday" animals as well as the "humans" with animal names and features that populate mythic time-space. In general, the *tsampi* and its denizens provide much of the content and color of the Cofán lifeworld.

It is important to note, however, that Cofán modes of relating to the *tsampi* do not exhibit a strong cosmological determination. Contemporary Cofán recount how old-time shamans "called" and "made" game animals through *yaje*-enabled communication with their spirit owners. Supernatural means of interacting with nonhumans, however, were never the predominant modes of human-animal relations. Moreover, Cofán people doubt that the game-calling powers of shamans were ever reliable.[1] Dream-based hunting augurs continue to be important, but the Cofán employ them for humor as much as hunting success. A man might wake up one morning and tell his wife, children, and anthropologist friend that he dreamt of making love to a fat woman. The symbolism means that he "dreamt of tapir." Often, he will laugh as he casts a teasing eye toward his wife. If he is about to head to his field to harvest plantains, he might take his shotgun, "just in case." The dream, however, is a sure sign of nothing.

In terms of gender, Cofán subsistence is relatively egalitarian. Both sexes garden, although men are more likely to clear primary forest for a new field, and women are more likely to harvest plantain, banana, and manioc. It is common for a husband, wife, and children to weed and harvest together. Moreover, men frequently go to gardens alone. Fishing occurs with almost no gender distinctions. Even hunting is less exclusively a male domain than it is among other Amazonians. Some women hunt with blowguns during their youth. Others go alone into the forest with a shotgun or a machete and a pack of dogs. Even in households where men procure all the meat, women often accompany their husbands or relatives in the forest. If an adolescent girl expresses interest, a male relative teaches her to hunt so that she can join him or hunt by herself.

If a man returns from a day alone in the forest with a good catch, his wife and daughters do most of the butchering. If it was a collective party, however, men and women butcher together. Women do the majority of cooking, although men either help or cook by themselves if their wives or daughters are otherwise occupied. If all of a household's women are under menstrual prohibitions, men do all the cooking and cleaning. More than anything else, menstrual taboos account for men's familiarity with women's domestic responsibilities.

Husband and wife butchering a monkey (Photo by Michael Cepek)

Village residents utilize borrowing, sharing, and giving networks for food and other items. If members of different domestic groups hunt together, they typically distribute the meat at the end of the outing. If someone uses a motorized canoe to fish or hunt at a distant location, people of all ages, genders, and households make use of the opportunity. If any member of the party is unsuccessful, they usually receive part of another person's catch. The likelihood of receiving game is higher if an individual contributed material goods to the venture (e.g., gasoline or a net), is from a household whose other members had no success, or is an older person who is no longer as skilled as he or she used to be.

Cooperative activities allow for socialization into a forest-based way of life. When children follow a relative on a hunting trail, the adult walks before them as a lesson in the forms of locomotion, gaze-shifting, and index-noting that are essential elements of embodied forest knowledge. As tracking and pursuit take hunters off well-worn paths, older people tell younger people what signs they are following and what outcomes they are intending. When hunters or fishers return to their homes, people who did not participate often become involved in the butchering process. While people investigate and comment on the physical state of the animals, hunting stories are told for the first time. Children and noninvolved parties watch and listen. While eating later in the evening, hunters replay the tales, especially if members of other households are visiting. In general, telling hunting stories is a relished form of entertainment and a good-natured means of gaining prestige. Older men get together on lazy afternoons to share banana drink and manioc beer while they laud each other's successes and joke about each other's failures.

Eventually, environmental knowledge moves from its source in primary, individual experience to the community's general stock of knowledge. Nevertheless, it would be an exaggeration to say that knowledge of Zábalo's forests is fully shared. Older men possess the most extensive knowledge of species' behavior and ecological processes. Active hunters, who are often younger and middle-aged men, are most aware of current population dynamics. Yet there is a fairly general awareness of the state of Zábalo's rivers and forests. Everyone plays a part in harvesting, processing, consuming, and commenting on the products of the *tsampi*.

Moreover, because of idiosyncratic interest or capacity, certain individuals know more than is expected of someone in their position. Although not an active hunter, the female head of an extended family in Zábalo knows far more A'ingae bird names than any other member of her household. Borman often tells outsiders that his wife is an "expert" in the nesting habits of birds. She apparently has a special interest in the topic, which she actively explores. My Zábalo ritual brother, while still in his twenties, had killed more tapir than almost anyone else in the village. He has worked hard to become an expert tracker, and he pursues the species for days at a time. Some men in Zábalo truly relish the hunt. Many women, in contrast, show fear of the deep forest. They prefer not to move far from the main settlement, either alone or with a party. Nevertheless, I know a young male household head who is not very interested in hunting. His wife procures almost all the meat that his family consumes. Without fear, she hunts and fishes on her own.

In general, Cofán subsistence allows for widely shared knowledge and the development of individual expertise. A person's environmental knowledge often escapes functionally weak norms of age, gender, and domestic affiliation. At the time of marriage, people expect men and women to be well on their way to knowing how to maintain a household of their own. Although the Cofán frown upon "lazy" individuals, there are many possibilities for intergender, intergenerational, and interhousehold aid.

Cofán openness to cooperation is matched by the impressive ecological competence of all individuals as well as the freedom to develop personalized ways of relating to the Amazonian environment. This sharing and openness means that people rarely have to act against their preferences to relate to the *tsampi* in an anxious or onerous way. The experiential benefit of interdependence and flexibility is a style of environmental practice that does not demand obedience to highly defined norms. Accordingly, the structure of knowledge and labor is an essential component of an *opatssi* life.

DORENO, TOURISM, AND CARING FOR THE *TSAMPI*

In the decades from 1970 to 2010, Cofán communities have become more stable and circumscribed. Andean and coastal colonists used oil roads to clear "unoccupied" land and expropriate Cofán territory. Cofán people still remember when they were "wanderers." Even middle-aged people recall traveling with friends and family for months at a time along the Aguarico and San Miguel Rivers. They hunted, fished, and harvested turtle eggs in a form of drifting use that they describe as *a'qquia injanga* (without structure, intent, or concern).

Before commercial oil extraction began in 1972, the Aguarico valley was sparsely populated. Its upper half was the home of only a few hundred Cofán. With such a low population density, overharvesting was not a concern.[2] After decades of observing large-scale ecological transformations, however, the Zábalo Cofán began to restructure their subsistence in an intentional way. They wanted to ensure the continued existence of the *tsampi* that remained within their community's borders. Their decision to *coiraye* (care for) their forest was the result of a number of developments. The most important factors were the experience of ecological alterations in Doreno and the insights gained by working with tourists.

Doreno residents gradually formed Zábalo by coming to the lower Aguarico with Western backpackers in the late 1970s and early 1980s.[3] It was not until the middle of the 1980s that the village became a primary residence site. Before Zábalo's emergence as a full-fledged community, Cofán people spent significant periods of time in Doreno. Many still do. In 1978, the community gained legal title to 9,451 hectares of *tsampi* on the south side of the Aguarico River. Today, Doreno's land is an island of forest in a sea of colonist clearings, oil fields, and agricultural plantations.

Both Zábalo and Doreno Cofán realize that life is now different in the latter's territory. In addition to recognizing the illnesses caused by oil wastes and colonist pesticides, people in Doreno eat much less fish, game, and fruit. It is virtually impossible to find materials for roofing houses or constructing canoes. When there is enough money to purchase them, canned tuna and sardines are the primary sources of protein. Hunger is a common complaint.

Doreno Cofán continue to hunt what they can. Nevertheless, they often spend entire days wandering their forest without encountering a single game species. Furthermore, Doreno animals are extremely *injansu* (suspicious). Compared to the *opa* animals living in less intensively used landscapes, they are difficult to kill. Many Doreno Cofán hunt mainly at night for species such as deer and paca, which have high reproductive rates and thrive in secondary forest. Hunting at night, however, is difficult and dangerous. Recently, one night hunter mistook another for a deer and killed him. If possible, Doreno Cofán hunt on forest fragments owned by neighboring colonists, who consider such animals as primates to be inedible. The practice depends, however, on an established friendship relation. If a Cocama catches a Cofán hunting on his land without permission, he might shoot him.

For Zábalo residents, observing the changing health and subsistence conditions at Doreno provides an object lesson on the consequences of life with little forest. Witnessing an ecologically transformed way of life provides a rationale for restructuring resource use in Zábalo. It raises awareness of the degree to which relatively sedentary subsistence can impact a circumscribed block of forest. In addition to the increasingly relevant distinction between *injansu* and *opatssi* game, there is a newly won knowledge of which species "will run out" and which "will not run out." Cofán people are now aware of their capacity to transform their environment in a way that makes it less suitable for the pursuit of everyday values, whose systemic absence threatens *opatssi* conviviality.

One could write an entire book about the history of tourism in Zábalo. The formation of the community and the growth of Cofán tourism went hand in hand. Nevertheless, by the time I began my dissertation research in January of 2001, tourism was virtually dead in the community. The U.S. State Department had issued a warning on travel in northeastern Ecuador because of the Colombian civil war. Three well-publicized kidnappings scared most tourists away, although small groups did make their way to Zábalo during my fieldwork. Their presence, however, did not compare to monthly group visits through the 1990s. The last holdout of Zábalo tourism was the Flotel Orellana, a floating hotel managed by the Metropolitan Touring Company. The boat stopped several times a week in the community for an afternoon lunch, a forest hike, and an opportunity to buy Cofán handicrafts. The people of Zábalo devised a way to distribute opportunities for income. Tourist money was the base of their cash supply through 2000.

Unlike tourism in other indigenous Amazonian communities, "ecotourism" in Zábalo focused on the forest and not on native life. The Cofán staged no ceremonies for visitors, who largely remained away from the central village in forest camps and huts on the Zábalo River. Certainly, elders' traditional dress and nose feathers fascinated the tourists. Nevertheless, older Cofán mainly related to tourists as forest guides. They used their environmental expertise to find the animals the tourists had come to see.

Borman was the main guide. Even though he spoke English without an accent and knew the language and logic of science, he was the central object of fascination. As with the other Cofán, he was comfortable with the interactions. Nevertheless, he always stressed that the Cofán were not objects of tourism but guides to what foreigners truly came to see. In an article in *Cultural Survival Quarterly*, Borman explains the Cofán philosophy:

> [T]he community as a whole decided early on that our role would be that of guides and service providers for tourists interested in the forest. We would not dress up and do dances, or stage fake festivals, or in any way try to sell our traditions and dress. We would not accept becoming the objects of tourism—rather, we would provide the skills and knowledge for the tourist to understand our environment. We would sell our education, at a price that was in line with its importance to the outsiders who wanted to buy it, such as lawyers, biologists, doctors, teachers, and other professionals the world over do. (1999:50)

The people of Zábalo know that tourists and scientists come to their village because of their maintenance of intact forest and their ecological knowledge. Without their animals or their abilities, Cofán realize that they would lose income, which is the entire motivation behind tourism, and to a similar extent, scientific collaboration. The importance of maintaining opportunities for tourism was a large aspect of their decision to restructure subsistence practices. In the words of one Zábalo elder:

> Before, before in the time when we did not make prohibitions on hunting, we came and we hunted a lot. A person would hunt white-lipped peccary one day. And the next day another person would hunt white-lipped peccary. And some days a person would kill woolly monkey, and another person would kill howler monkey, and another person would go and kill woolly monkey, and another person would kill macaw. Really, at that time there was little left on the Zábalo River. You could barely see a macaw! If it becomes like that, and the tourists come and we guide them, what will we show them? So that began to happen and we thought, "We should stop acting like that."

Aside from the instrumental equation of forest with cash, a more paradigmatic transformation took place during the move from Doreno and the growth of tourism. People in Zábalo say that they prefer tourist work because it is "not damaging." They realize that virtually every other mode of making money—oil extraction; ranching; farming cash crops; mining; and selling meat, hides, and other wildlife products—causes harm to the forest and its nonhuman inhabitants.

Zábalo residents spent many hours in intimate circumstances with tourists. They describe tourists' encounters with animals by mimicking their smiles, laughter, and excitement. Cofán guides recount how tourists become happy when they find one of their "desired" species. They do not disparage tourists for their fascination. Rather, they conceive of it in the same way that they imagine themselves staring at American skyscrapers: inherently fascinating objects that hold one's eyes because they have never been seen or even imagined before.

By working with tourists, Cofán people learned a different way of looking at the forest. This new orientation, which we could describe as "appreciating without directly using," became part of a broader relationship between the people of Zábalo and their *tsampi*. Now more than ever, the Cofán appreciate the autonomy of the forest and its nonhuman inhabitants. In Borman's words:

> Through our dealings with tourists, we had become aware of a whole new dimension concerning the value of wildlife. To us, a woolly monkey was something that tasted good. All of our relations with the woolly were directed

toward putting it in the pot. This didn't mean we didn't enjoy its antics or observe its behavior—on the contrary, we needed to know a tremendous amount about woolly monkeys to be able to put them in that pot. But to observe them for observation's sake, for their aesthetic value alive and in the wild, just did not occur to us until we began dealing with tourism.

The sense of autonomy has a double valence. First, although Cofán people did not have a previous attitude of total control over the *tsampi*, the guiding orientation behind encounters with nonhumans was pragmatic. People hunted what they found, when they found it. Letting something be, or just watching and observing it, was not a common stance. Second, witnessing changes in Doreno and becoming involved with tourism taught the people of Zábalo to relate to the *tsampi* as a delimited object. They began to understand it as something that might or might not exist because of human, Cofán agency. In other words, rather than being the medium of all life, the *tsampi* and its denizens became detached from their everyday being as the desired outcomes of hunting and harvesting. They assumed the shape of something that could stand apart from human action because of human action. Gradually, the Cofán began to understand their environmental practices according to an opposition between activities that "damage" and activities that "care for" the *tsampi*.

The words of a Zábalo elder express many of the relevant dynamics: opposing a forest-based way of life to a cleared land–based way of life, interacting with tourists as a means of reimagining the *tsampi*, and realizing that Cofán actions produce environmental changes:

> If we care for the *tsampi*, the animals will live. Collared peccary, woolly monkey, trumpeter, everything. They will live, and sometimes we'll kill them and eat them, but they will live. On the other hand, one could make pastures here. Five hectares, fifty hectares, all the way to the Daroé stream. But then where would the woolly monkey live? There wouldn't be any. They would disappear. In order for that not to happen, we began to think, "Yes, let's really care for our land."

COMMUNITY SUBSISTENCE REGULATIONS

Caring for the *tsampi* has become a community-wide institution and an everyday practice. By the time Zábalo became part of the Cuyabeno Wildlife Reserve in 1991, the Cofán had begun to formalize a system of controls on their subsistence practices. The transformation added a layer of rules to their long-standing environmental orientations. The most important aspect of becoming *tsampima coirasundeccu* (caretakers of the forest) was the creation of communal *se'pi'cho* (prohibitions).

Zábalo Cofán remember the mid-1980s as the time when environmental prohibitions began. Most residents explain the creation of conservation rules as a natural extension of their desire to avoid the fate of Doreno by caring for their forest in a more structured way. Although Borman's leadership and the extension of the Cuyabeno Wildlife Reserve were important parts of the process, Cofán people describe the formalization of *se'pi'cho* as a collective, organic, and community-internal process. All Zábalo residents now accept, participate in, and understand the practice. After more than two decades of experimentation with limiting subsistence, a broad system of prohibitions covers a diverse array of species and activities. The rules fall into six categories:

> *Prohibited species*: Because of their delicate ecological state and/or interest to tourists, certain animals are never killed. The list includes macaws, raptors, caimans, river turtles, otters, large felines (except when threatening humans), giant armadillos, anteaters, dolphins, anacondas, hoatzins, anhingas, herons, sun bitterns, sun grebes, and ibises.[4]

> *Prohibited areas*: Zábalo has a "zoning" system. The structure has three parts: a small area for intensive use (dwelling construction, gardening, lumbering, fishing, and hunting) along the main body of the Aguarico; a larger area for extensive use (limited hunting, fishing, and harvesting of some plant species) beyond the intensive area; and an even larger reserve area (for tourism, scientific research, and animal reproduction) in areas far from human habitation. The great majority of Zábalo territory is "reserve." The system also has finer divisions. For example, people do not hunt certain species, such as primates and curassow, along the Zábalo River or the south side of the Aguarico. The division between hunting and no-hunting areas is lexicalized in A'ingae as that between *fi'tti'je'cho* (hunting places) and *atapa'je'cho* (places of animal reproduction). In addition, for reasons of tourism, no one cuts trees along the shores of the Zábalo River or paths where tourists walk.

> *Limited takes*: There are a number of species that can be killed in limited amounts. Some, such as piping guan and woolly monkey, are limited according to calendar year (three each per household per year). Others, such as white-lipped peccary, are limited on a per-outing, per-hunting-party basis. (Each party can kill two peccaries per trip.) Moreover, household members cannot kill additional white-lipped peccaries if they still have smoked meat from a previous trip. Consequently, there is a temporal spacing of hunting trips.

Limited seasons: Some species can only be hunted during months when they are fattiest. The logic is to avoid waste by killing individuals only when they are largest and best in taste. People kill woolly monkeys between May and August and piping guans in October.

Prohibited reproductive states: Before instituting a total ban on the species, people in Zábalo prohibited the hunting of juvenile caimans and the capture of egg-laying river turtles.

Prohibited activities: There are a number of activities that are totally prohibited in Zábalo. The list includes commercialization of almost all forest products, including meat, hides, live animals, turtle eggs, lumber, and dugout canoes. In addition, fishing with dynamite or artificial poison is prohibited. The use of natural fish poisons is allowed only in ponds left at the end of the flood season and never in moving water.

The concept of *se'pi'cho* is important in many areas of Cofán culture. In its most general sense, a *se'pi'cho* is a restriction that an agent makes to prevent another's engagement in an activity or use of an object. Common *se'pi'cho* include the menstrual prohibitions, which restrict a woman's routines during her menses. Whereas people refer to menstrual prohibitions as *se'pi'cho*, they do not attribute their creation to individual humans. Rationales for following them, however, include fear of recriminations by supernatural agents, such as the *yaje a'i* and *cocoya*, whom menstrual blood disgusts and angers.

Ritual experts of both sexes declare many of the most difficult prohibitions. When counseling a person through the unstable states associated with illness, menarche, male adolescence, shamanic training, pregnancy, and the postpartum period, specialists advise certain forms of avoidance, especially of particular foods. Human actors clearly make these *se'pi'cho*. They usually explain the prohibitions according to their understanding of the amount of blood in the relevant animals, which can trigger uncontrollable loss of blood and life force in fragile individuals. They also cite an iconic resemblance between the animals and the physiological states to be avoided.[5] Although the prohibitions are based on shared medicinal and cosmological principles, there is substantial latitude in determining whether they apply to specific individuals. Accordingly, people say that named experts explicitly prohibit the relevant foods and activities.

Se'pi'cho on environmental activities are different. They are products of community-wide discussions, which culminate in formal decisions at Zábalo's end-of-the-year meeting. All environmental *se'pi'cho* are explicit restrictions on normally pursued activities and frequently used objects. In other words, they are rules. Nevertheless, because they achieve written form, apply equally to all community members, and are backed by enforcement practices, they function as "laws" in the classical meaning of the term.

The community president calls and directs the December meeting by "gathering up" residents and acting as the main speaker. After dealing with other issues, the elected secretary announces the list of existing prohibitions, which he keeps in written form in a community folder. If someone expresses a desire, they begin to discuss changing the rules or creating new ones. When residents share their opinions on making a *se'pi'cho*, they support their positions with reflections from daily subsistence. Both men and women chime in with corroborating or conflicting observations. Slowly, a consensus begins to emerge. If there is little or no opposition, the matter comes to a "vote." The president asks every resident whether they will abide by the new prohibition. If everyone answers yes, the secretary enters the prohibition into the community folder along with the existing regulations. Then, people say that they have *ñoñan'fa* (the plural form of *ñoñañe*, the verb for making any artifact) the *se'pi'cho*.

A few examples serve to describe the process. In the first account, a resident explains the discussion of whether or not to prohibit the killing of an animal. He also talks about the common practice of initiating a prohibition with a

Members of three households and two generations hunting together (Photo by Michael Cepek)

one-year period of increased attention to discover whether the prohibition is necessary, whether the population is changing, and whether the restriction was effective:

> Everyone will speak. "Come on, what do you all think? Should we stop hunting this and watch to see what happens?" That's what the president asks everyone during the meeting in December. And people will say, "Yes, yes, yes." Everyone. He asks everyone, "Will you do it like this? Yes? Will you do this?" One person, another person, everyone. Man, woman, man, woman, everyone. All community members. You must do this as a *comunero*. If everyone says "yes," *ya*, then it will be written that that animal is *fi'tti'masia* (unkillable). That's how we make these things.

Sometimes, a person discusses a possible prohibition, but there is not enough agreement to bring the matter to a vote:

> [Who speaks first about the possibility of prohibiting something?—M. Cepek] Whoever. Whoever wants to. If someone wants to, they will speak and it will go out there. And then all people will think, "Yes, you spoke truly. I, too, thought like that but didn't speak." Then the discussion will start. I want to prohibit tapir. Before, you would see many tapir footprints along the Zábalo River. But now, nothing. Now you go far upriver and you'll see tapir tracks. Maybe one, somewhere else one, where they come down to the river. Some people, though, will say, "They won't run out. How could tapir run out? The *tsampi* is big." But we know their tracks! Before, there were so many tracks here, right at the mouth of the river. Now there are none. Seeing that, people say, "They are beginning to run out." But other people say, "No, they are not beginning to run out."

Often, a spirited discussion emerges about whether or not to prohibit. There is an unspoken rule that if the people who want to prohibit far outnumber the individuals who do not, the majority will "win," and the people who oppose the measure will relent. As one person explained:

> Some people will want to prohibit. A lot of them, let's say ten people. And only one person won't want to, saying, "Just let it be. Let's stop this. Don't make that prohibition." That person wants to hunt the animal. He says that, and the others say, "Think about your children. You're a parent. What will your children eat in the future if their parents finish off everything?" That person will then sit and think. "Yes, all right, that's fine. All right, let's go ahead, and then we'll know. Will the animal really run out?" He says that,

and has the animal really begun to disappear? Yes. "It has begun to run out," people will say. Then that person will think, "Yes, it's true, it just recently happened." He went to the forest, and he came back with nothing. The next day he went to the forest and he found nothing again. He killed nothing, and he knew that the prohibition was justified.

Multiple residents expressed the same understanding of how power structures the discussion and enactment of prohibitions. Everyone I interviewed voiced a basic theory of majority-rules democracy. If a person voices his or her concerns at the December meeting, the topic enters collective debate. Everyone has a "vote" on whether or not to declare or change a *se'pi'cho*. People stress that prohibitions are more than the doings of a *na'su*, be it Borman or anyone else. When I asked one middle-aged man if the president can "command" people not to kill something, he replied:

> The president does command. If people say "yes," if that's what they want, the president will say, "Have you all listened and understood? That animal is unkillable."[6] That's how the president scolds and councils everyone. [So is the *se'pi'cho* the president's command, or everyone's doing?—M. Cepek] It's the command of all people, everyone. How would the president be able to do it? If the president says, "Do that,"[7] people will say, "*Tsss*, no, forget it." How could he do it? It's like that. All people want something, and the president, too, has to say "yes." He does it, and then it's written and left, "It's bad to kill that animal, and there is a fine of $12 for doing it."

The last statement identifies one of the central components of the *se'pi'cho* system: the publicly demanded *multa* (Spanish for "fine," also referred to as *afe'poeña'cho* [obligation to pay] or *shaca* [debt] in A'ingae) for breaking a rule. People do not remember how they decided the exact amount. They do, however, recall its shifts, especially when Ecuador's national currency changed from the sucre to the U.S. dollar. Everyone agrees that the amount is far more than anyone would like to pay. Nevertheless, it is within reach of most families.

The individuals I interviewed were willing to discuss their own and others' infractions. Examples included: killing four white-lipped peccaries instead of two on one outing, crossing the Daroé stream (a prohibited area) to hunt woolly monkey and Spix's guan, and killing a hoatzin out of frustration with one's inability to shoot other animals. Rule breaking becomes a public fact. Infractions follow a route from immediate admission or covert discovery, to informal gossip channels, and finally to open acknowledgment in the December meeting. The *na'su* publicly questions the accused individual, who publicly responds, usually with an admission. In addition to the burden of paying the fine, open admonishment is a real embarrassment. It is a fate that all people want to avoid. Furthermore, subsequent fines for the same infraction double each time, escalating far beyond the means of many families to pay them.[8] Once a person has orally agreed to abide by a prohibition, there is no escape from paying the fine and being embarrassed in front of the entire community.

Zábalo's *se'pi'cho* system represents an important shift from earlier modes of action. Above, I introduced the phrase *a'qquia injanga* (without structure, intent, or concern). People use the words to describe any activity that is not encumbered by preoccupation or excessive attendance to external concerns. Conceptually, *a'qquia injanga* is related to the term *opatssi*, which Cofán understand as freedom from anxiety or fear. Hunting and fishing in an *a'qquia injanga* way means losing oneself in the immediacy of the action. An individual has no concern for either her personal safety or the larger social or ecological consequences of her action. Currently, Zábalo residents consider *a'qquia injanga* subsistence to be detrimental to the *tsampi*. Consider one man's account of how he hunted when he first came to Zábalo:

> I killed and ate a lot of macaws when I first came here. Before the time of prohibitions. Whatever you wanted to shoot, you just shot it! Everything. Piping guan, macaw, white-lipped peccary, howler monkey. Everything that you could shoot. If you could actually hit it, why wouldn't you just shoot it? With no good reason. Even if you still had a lot of smoked meat, you'd really want to go out and shoot some more. Truly, that's how I was. And that's how I damaged too much before.

By creating and following prohibitions, Zábalo residents do not subsist in an *a'qquia injanga* fashion. Consequently, they do not consider their practices to be excessive or damaging. One young man said, "Our lifestyle is truly not excessive.[9] That's our lifestyle. That's why we have our limits. So that, for the future, we don't hunt too much." If you ask someone if ethnic others know to prohibit game, the most common response is *a'qquia injanga fi'tti'je'fa* (without structure, intent, or concern they hunt).

Subsisting with care, however, means bringing a new set of concerns into daily life. On Zábalo hunting trips, I have seen the experiential difficulties of restraining oneself from taking a shot at a prohibited animal. The restriction amounts to a frustrating stifling of instinct. On one occasion, I ran with hunters and chased a troop of woolly monkeys to the Daroé stream, where the no-hunting area begins. The party cursed its luck and waited for the animals to return from the other side, which they failed to do.

At an even deeper level, living with prohibitions means controlling one's emotions. A young man killed a hoatzin after he repeatedly missed the fish he was trying to hit with a .22 rifle. He had no intention of eating the hoatzin, an "inedible" bird, yet he had to pay a fine anyway. In addition, people must be much more cautious with their shooting. When a man took two shots with a 20-gauge shotgun and brought down four white-lipped peccaries instead of the allowed two, he had to pay the fine. His infraction was not the fault of intention but of an unintended spray of shot into a herd of fleeing animals.

Importantly, game prohibitions also disrupt sociality. When relatives from other communities visit, they pertain to the household with whom they stay. If the untutored visitors hunt a prohibited species or kill in excess of a specified limit, the head of the household that hosts them pays the fine. Even if Zábalo residents adequately explain prohibitions to visiting kin, it is difficult to communicate up-to-date information on hunting takes and stores of smoked meat. For example, a man killed two white-lipped peccaries one day, and his visiting cousin killed two peccaries the following day. The community judged the take to be four peccaries in one household at one time, which exceeded the limit. As a result, the Zábalo resident paid a fine.

Living with environmental prohibitions is burdensome. Dislike for prohibitions can be so extreme that some Cofán cite the game laws as their main reason for not moving to Zábalo. One old Doreno resident has numerous kin in Zábalo. He would definitely appreciate the *opatssi* quality of life in an area far from colonists and contamination. When I asked why he does not move to the community, he gave one answer: "Too many prohibitions!"

Environmental *se'pi'cho* represent the creation of a new sphere of concern and control as well as a new mode of knowledge production. The prohibitions allow for the collective restriction of action in order to produce effects on a total object: Zábalo's *tsampi* in its entirety. Individual knowledge and agency become community knowledge and agency as Zábalo residents collectively discuss, control, and learn from the consequences of their actions. The process allows for the confirmation of the initial decision, its refutation and subsequent abandonment, or its modification according to observed changes. People often refer to prohibitions in an experimental light, as "observing without hunting," "observing without using," or "observing after halting an action."

Living with *se'pi'cho* does not replace earlier forms of producing and circulating knowledge. In fact, creating prohibitions depends on them. The openness, sharing, and mutuality of Cofán environmental relations make the collective creation of *se'pi'cho* possible and compelling. The people of Zábalo find no reason to doubt the efficacy of the institution or the knowledge it creates. Even though they recognize how burdensome the *se'pi'cho* system can be, they have seen its possibilities for extending desired ecological states into the future. They claim that as long as they control their territory, they will not abandon their practice of *tsampima coiraye*.

SCIENTIFIC CONSERVATION

Cofán conservation arose in relation to a number of non-Cofán actors and forces: ecologically oriented Western tourists, a distant state environmental ministry, and the destruction caused by the oil industry and large-scale colonization. Nevertheless, the people of Zábalo developed their practice of *tsampima coiraye* by themselves and for themselves. They saw little need to communicate its structure or its consequences to outside audiences. Two groups of newcomers, however, have shown a keen interest in Zábalo's forest and Cofán action: Western scientists and conservationist NGOs. Zábalo's location inside a protected area intensifies involvement with the groups, who have been working in the Cuyabeno Wildlife Reserve for decades.

Like the Cofán, scientists and conservationists want to keep Amazonian Ecuador ecologically intact. Nevertheless, Cofán people practice their own system of environmental care. Although outsiders can see and study the plants and animals of Zábalo's forest, the encounter need not involve familiarity with Cofán knowledge. The circulation of ecological information in Zábalo is based on social intimacy and cultural similarity. It does not involve the processes of standardization, formalization, and quantification that have made portability, objectivity, and truth mutually implied terms in contemporary scientific practice (Porter 1999).

Although individuals in Zábalo have guided and hosted dozens of researchers, their most important encounter with Western science and conservation was their partnership with the Environmental Conservation Programs (ECP) of the Field Museum of Natural History.[10] Although the ECP continues to support Cofán conservation initiatives on multiple levels, the majority of its work in Zábalo occurred between 1999 and 2005. Founded in the mid-1990s, the ECP is the wing of the Field Museum that works on "conservation action," as summarized in its introductory Web-page statement:

> Environmental and Conservation Programs (ECP) was established in 1994 to direct The Field Museum's collections, scientific research, and educational resources to the immediate needs of conservation at local, national, and international levels. ECP is the branch of the Museum fully dedicated to translating science into action that creates and supports lasting conservation.

Through partnerships with research institutions, conservation organizations, local communities, and government agencies, ECP catalyzes science-based action for conservation. (Field Museum 2005)

The ECP focuses most of its work on the Amazonian regions of Ecuador, Peru, and Bolivia. Debby Moskovits, the founder and head of the ECP, learned of Randy Borman and the Cofán of Zábalo through publicity of their opposition to oil development and their work in community-based conservation. She subsequently helped Borman receive the Field Museum's prestigious Parker/Gentry Award for Conservation Biology in 1998.

For the better part of a decade, the ECP developed environment-centered relations with the people of Zábalo. Multiple objectives oriented their involvement with the community. First, the ECP wanted to document what the Cofán were doing prior to its entrance in order to show the world that indigenous conservationism is possible. They believe that Zábalo's practice of *tsampima coiraye* represents a compelling reason for indigenous-environmentalist alliances. They explicitly oppose their vision to the position of other large conservationist NGOs, some of which have abandoned collaborations with indigenous peoples due to direct questioning of the myth of the "ecologically noble savage." And second, the ECP sought to incorporate scientific logic, methods, and instruments into Cofán practice. Between 1999 and 2005, they labored intensively to make science an integral aspect of life in Zábalo. They believed that by becoming scientists themselves, the Cofán would be able to manage their resources in a more rational, efficient, and effective way. In effect, the ECP wanted to provide the Cofán with better tools for pursuing their long-standing objective of forest care.

ECP personnel traveled to Zábalo multiple times a year. During their stays, they taught the use of technological instruments (e.g., notebooks, computers, and measuring devices), they created a basic infrastructure (e.g., a project meeting house and a system of census trails), and they communicated the utility of scientific forms of understanding. Eight Zábalo residents worked on ECP-supported projects, including two coordinators and six monitors. All of the workers pertained to different households and earned a monthly paycheck.[11] Of the following activities, the ECP modified the first two and initiated the next three:[12]

1. *Beach monitoring of river turtle nests*: This project involved watching Zábalo's beaches for the laying of eggs by two endangered species of river turtle, *Podocnemis expansa* and *Podocnemis unifilis*. During the egg-laying season, which runs from October through January, community members

searched the beaches. When they found a nest, they either marked it with sticks or harvested it. If they marked it, they told a monitor, who went to the location and labeled the nest with a number, a date, and the name of the person who found it. They also told monitors if they harvested and consumed eggs. Monitors then communicated the information to a coordinator, who kept a computerized record. Each monitor had to watch the beaches on his assigned stretch of the river, which was divided into five sections. If the river flooded, the eggs were lost. If there were advance signs of flooding, the monitor moved the eggs to an artificial beach above the flood level. If the eggs hatched successfully, the community member who marked the nest received $.25 per hatchling for *Podocnemis unifilis* and $.50 for *Podocnemis expansa*, which is a much more endangered species.

2.   *Head-starting of river turtles*: The project coordinators collected turtle hatchlings from their nests and put them into artificial ponds, which they maintained next to their homes. They fed the hatchlings every day, and they maintained records of air and water temperatures. They weighed and measured the hatchlings three times: when they received them, at six months, and at one year. After a year in the pools, the hatchlings were released into local lakes, where they grew before dispersing into rivers. The logic behind this project, as well as the beach-monitoring project, was to grow the hatchlings under protected conditions in order to increase their survivorship rate. The ultimate objective was to recuperate the population in the Aguarico River, which has suffered centuries of commercial egg harvesting.

3.   *Visual census of river turtles*: The objective of this activity was to track turtle populations in Zábalo's territory. Five monitors canoed along divided areas of the Aguarico and Zábalo Rivers, as well as a number of lakes, to make a visual count of juvenile and adult turtles. They indicated the location of the turtles according to a system of plaques that were placed every 250 meters along the rivers. Monitors classified the turtles as small, medium, or large. In addition, monitors kept records of climatic conditions and river levels on their census days. Every month, each monitor completed two one-day censuses on the Aguarico River and one two-day census on the Zábalo River. After collecting the information, they gave it to the coordinators, who entered it into the project laptop on an Excel spreadsheet.

4.   *Terrestrial census*: The objective of this activity was to observe game animal populations and to discover population fluctuations in varying habitats, including hunting and no-hunting zones. The monitors walked on

trails to record sightings of terrestrial, arboreal, and avian species. Every month, they completed two one-day walks of five kilometers and one two-day walk of twenty kilometers. The long trail was half inside a hunting area and half inside a no-hunting area. The monitors constructed and maintained the trails, placing distance-markers every fifty meters to locate sightings. They took notes on a number of variables: date, time of departure, time of sighting, species sighted, number in group, mode of observation (direct sight, sound, or footprint/marking), distance from trail, and location. The coordinators entered the data into an Excel spreadsheet.

5. *Hunting tabulation*: The objective of this activity was to create a record of Zábalo hunting. The monitors questioned all community members and recorded their takes. They wrote the information in a notebook, specifying name of hunter; location of hunt; date; and the species, number, sex, and reproductive state of the hunted animal. The monitors then turned in the data to the coordinators, who entered it into an Excel database.

All of the ECP-supported work involved structuring human activity and the physical landscape. Spatially, workers segmented the rivers in 250-meter sections for visual census work and in five broad sections for beach-monitoring work. They divided terrestrial census trails into 50-meter segments. Temporally, workers coordinated project activities according to basic divisions of the calendar year, structuring work cycles both annually and monthly. Practically, workers moved, collected, constructed, and perceived in uniform ways. Without one standard mode of action, it would have been impossible to compare the results of different monitors or to ensure that turtle hatchlings received equal treatment.

All of these activities depended on the creation of texts in written A'ingae and the standard Western number system, which enabled their incorporation into computerized databases. The ECP struggled for years to make sure that workers used the same species names, spelled in the same ways. Workers recorded turtle sightings according to a size system—small (<20 cm), medium (20–30 cm), and large (>30 cm)—that has no correlate in A'ingae. For the hunting tabulation, the ECP struggled to help the Cofán create a system of numbered and standardized place-names that monitors could use to pinpoint hunting locations on an "accurate" map. Finally, the work demanded maintenance of a set of records: weights and measurements of turtle hatchlings; "owners," dates, and locations of marked nests; and air and water temperatures, as well as feeding details, for the turtle ponds.

The new forms of moving, sensing, and writing created a "picture" of Zábalo's environment: the turtles of its rivers, the animals of its forest, and the human harvesting of each. On a single sheet of paper, the results of thousands of hours of activity could be reduced, for example, to a table of game animal distributions in hunting versus no-hunting areas. Quantified, uniform, and synoptic, these scientific representations allowed for the discernment of patterns with a level of "exactness" that neither everyday subsistence nor community discussion could provide.

For the ECP, rigorous data collection and exact computation and representation presented a decisive advantage over preexisting modes of environmental knowledge. Participants put all the information into textual forms, which displayed how absolute numbers (e.g., of turtle nests protected, peccaries killed, or toucans seen) varied according to calendar year as well as spatial location (e.g., of marked-off river section, numbered hunting location, or segmented census trail). And when absolute quantities or computed averages did not suffice to communicate a trend, people could place totals of varying kinds on a chart in chronological order, allowing the ups and downs of a single line to tell the years-old story of an entire territory.

Rather than culture- and context-bound discussions about what kinds of animals are either "running out" or "increasing again," the ECP's program of scientific conservation produced an entirely different kind of object: stable, portable images with colors, numbers, names, and maps. The intent was to have the representations speak in the same way about the same things to a Western scientist, a government official, an oil company executive, or a Cofán leader. From the ECP's perspective, scientific information allows any individual to be as knowledgeable as any other. It depends on neither the memory nor the objectivity of particular, isolated individuals. Nor is it restricted to the localized trust and sociodiscursive confusion of annual community discussions. For Cofán people, however, neither the work required to produce scientific representations nor their overall utility are convincing in the same way.

On a few occasions, project participants engaged ECP personnel in direct conversations about the logic behind their interventions. In one interaction, a Cofán worker questioned an ECP staff member about the meaning of the term *evidencia*. The scientist explained the word with an opposition between census-produced *datos* (Spanish for "data") and *información*, which are *números* and therefore *más preciso*, and *opiniones*, vague senses about which there is not always *consenso*. The opposition between fact and opinion is highlighted in a Spanish-A'ingae pamphlet that the ECP developed to document Zábalo's experience with the terrestrial census. A question is posed at the beginning: "Why do we do a census program of the animals in our community?" Then, an answer is suggested: "A terrestrial census program can help us to obtain information for making decisions." One page is dedicated entirely to communicating the fallibility of "opinion," with different Cofán individuals pictured as wondering to themselves: "I want to know how things truly are. What should I believe?" The booklet describes the techniques used in Zábalo, with a running commentary on the utility of science: "There are ways of finding answers—doing regular censuses. Doing regular censuses can help us to discover tendencies. Scientists use methods like this in order to test their ideas and to know what's happening—and we can do it, too." The publication portrays the desired endpoint with an image of a Cofán man and woman pointing to diagrams, numbers, maps, and papers with ten years of census results in their hands. In conclusion, it reads: "Now we base our decisions on real information, not on opinions."

The Cofán experienced scientific conservation according to long-standing social and environmental orientations. Workers referred to their project participation as *semamba*, a nominalization of the verb *semañe* (to work). The word refers to sustained manual activity (e.g., making a field or building a house) and all forms of wage labor.[13] Even though project activities took monitors through the same spaces that they traverse while hunting and gathering, the moment-to-moment flow of their action differed radically from that of daily subsistence.

For example, whereas an individual might hunt on a census trail, terrestrial monitoring demanded unique forms of sensing and moving in the same territory. Monitors stayed on the trail at all times. Distance markers every fifty meters functioned as cues to stop, to listen, and to watch. When workers detected an animal, they paused to take out their notebook and record the encounter. The terrestrial census differed from the starts, stops, zigzags, and varying durations of hunting trips, which demand attention to traces and bodies rather than pen, paper, and mental distance calculations.

The monitors internalized the "scolding" of ECP personnel. The scientists told them to walk slowly, to stop at random moments to watch for wildlife, and to reproduce the same structure of action on each outing. Additionally, ECP staff decided that monitors should not carry shotguns or rifles while doing census work. They wanted to distinguish the activity from hunting and to guard against the temptation of taking a shot at an encountered animal. Walking in the forest without a gun, however, arouses deep fear in some monitors, as Cofán people are perpetually wary of jaguar attacks.[14]

In short, although terrestrial census work depended on preexisting environmental knowledge, monitors oriented the knowledge to novel ends in an activity that entailed new forms of preoccupation. The demands represented a multiplication of the concerns of *se'pi'cho*-structured hunting. Considering the difficulties, monitors qualified the work as an intense *vanan'cho* (struggle or suffering).

Other aspects of ECP projects were equally challenging. Wielding measuring instruments was a new skill, and people were afraid of damaging expensive high-tech objects such as binoculars, laptop computers, and GPS locators. Accidentally destroying them would have disrupted work and, many assumed, angered ECP personnel. For most of the monitors, writing and reading were difficult tasks. At least two of the older workers depended on the aid of their school-age children. The more knowledgeable coordinators repeatedly corrected the monitors. Monitor data sheets became public objects in project meetings, which allowed for explicit competition and critique. The interactions led to conflict, tension, and the painful awareness of failure and inability.

Project sociality was different from everyday life. Some workers found it easy to laugh off the difficulties. Others showed intense discomfort with the corrections of the coordinators, who sometimes referred to themselves as "bosses" and the monitors as "children." Monitors performed identical actions, which produced visible products (i.e., written numbers and words) that allowed individual capacities to be measured against one another in a public framework. Cofán people prefer to avoid situations structured by power differentials, evident inequality, and open criticism, which were necessary components of ECP activities.

The relatively strict scheduling of project activities made them even more onerous. When turtle eggs were not being laid or hatched, workers performed nine to ten days of labor a month: three for the terrestrial census, three for the turtle visual census, and three to four days to gather the hunting information, to write up the data, and to meet with other workers. Typically, the Cofán pursue subsistence activities according to their shifting tastes, energies, and interests. In contrast, project workers had to meet monthly requirements and collectively coordinate their actions. During the egg-laying season, they were in Zábalo every day, as community members constantly reported newly marked nests. In addition, monitors had to guard against flooding, which sometimes woke them up in the middle of the night to transplant threatened eggs. To satisfy the requirements, project workers could not leave the community during half the year. Consequently, the important practices of visiting kin and friends in other villages, as well as buying commodities in Lago Agrio, became impossible.

The structure of ECP-supported activities created tensions between project workers and the rest of the community. Community members participated in the turtle project by searching for new nests and either harvesting the eggs for consumption or "selling" them to the project for a small price (i.e., $.25 for *P. unifilis* and $.50 for *P. expansa*). They only received the money if the hatchlings survived to be placed in a pool. For many people, the revenue was significant, approaching $100 a season. If monitors failed to move the nests in times of flooding, the reaction of nest owners was extreme.

One year, dozens of nests vanished when a group of monitors risked a trip to Lago Agrio during the egg-laying season. No one moved the nests before flooding occurred, and the individuals who marked them lost the expected income. The incident led to a severe strain on the project. Some Zábalo residents suggested dismantling the monitoring structure so that all community members could guard their own nests and receive part of the workers' income.[15] The following year, many individuals chose to eat the eggs rather than contribute to the project. Their decision was part rebellion and part rational choice, as they intended to get at least some satisfaction for their labor.

Other Zábalo residents expressed a more general dissatisfaction with the project structure. For six years, the same individuals worked with the ECP. As other community members lost work opportunities or came of age, there was increasing resentment that some residents had a regular income while others had nothing. There is an intense leveling dynamic in Cofán society. People who feel as if everyone is not equal will "become angry" and "become hateful," thereby taking on the role of "criers." Dealing with the jealousy, resentment, and bad talk presented a serious problem. Only in the context of broader community critiques did workers voice a desire to quit the project altogether. They wondered whether $100 a month was worth the ill will that their participation inspired.

Relations with ECP personnel could also be stressful. Project workers viewed the scientists as their bosses. Visits were particularly difficult for the coordinators, who felt that they were responsible for the work of the monitors. Some ECP personnel were gentler than others when dealing with the workers. Nevertheless, the Cofán became tense when the scientists asked questions such as: Why isn't all of the data in the computer? Why doesn't a monitor know how long his census trail is? Where are the receipts for the gasoline that was bought to fuel the project generator and water pump?

The workers feared the recriminations. The power structure made them keenly aware of whether visiting ECP personnel were happy or angry with their performance. Tracing the emotional reactions of ECP personnel became the central means of gauging one's own performance and overall project success. In general, Cofán people do not like to be criticized in front of their peers. One's inability to perform magnified the discomfort, as the workers feared that their failure meant the disappearance of their income. The most anxious workers drank an extra bowl of manioc beer before meeting with the *museo* representatives, which helped to calm their nerves.

In short, participation in ECP-supported activities represented a substantial transformation of the Cofán lifeworld. New objects became part of new activity structures and new institutions. Their performance entailed new forms of exertion, preoccupation, anxiety, and conflict. Project-related struggles meant that scientific conservation was much more than the intuitively useful or enjoyable activity that some ECP personnel believed it to be.

The main clash with the ECP's mission, however, was Cofán people's inability to understand how scientific knowledge could help them to *tsampima coiraye*. Absolute numbers, statistical averages, texts, maps, tables, and diagrams are representational forms beyond the reach of most Zábalo residents. In June of 2004, I was present at the first community meeting in which ECP personnel shared data from the terrestrial census. The main problem was that virtually no one understood the concept of "average," which was central to two types of results: "distance rates" (kilometers walked to see a specific species) and "encounter rates" (number of animals seen for every ten kilometers walked, taken out to two decimal points). Although it is a much more familiar practice, even the *se'pi'cho* system is a relatively recent development, and community residents did not recognize its utility overnight. Moreover, many of them still experience its relatively minimal transformations of daily practice as a burden. Nevertheless, Cofán people find the knowledge that it produces convincing. Its existence does not depend on the representational forms and troubling labor that are central to science.

Overall, there are a number of explanations for Cofán people's failure to internalize the logic of scientific conservation: they experienced the work as difficult, burdensome, and surrounded by conflict; workers feared ECP personnel as their bosses; and neither project participants nor the rest of the community understood how the practices could help them maintain the ecological states they value. An additional explanation is also important: Zábalo residents found it difficult to imagine that they could ever possess enough resources to sustain the projects on their own, given the expensive, exotic equipment on which the work depended.

In general, Cofán people see little community-internal use for the numbers, words, and charts that the ECP projects produce. Like historian of accounting Theodore Porter (1999), they understand formalization and quantification to be most useful in transporting knowledge across, rather than within, established lines of culture and power. Accordingly, they deny science's ability to improve community-internal practices. Instead, they believe that science transforms their knowledge into a good that can be understood and used by non-Cofán outsiders.

Although the Cofán have a long history of peaceful interethnic cooperation, they are sick to death of five hundred years of unequal relations with Westerners. In the twenty-first century, they do not believe that any Westerner—environmentalist, humanitarian, or other—would come to their communities if not motivated by self-interest and profit. The Cofán maintain a strong set of expectations for such encounters: that outsiders will attempt to make money off them, that the most altruistic foreigners see them as objects of missionary-like charity, and that their only sensible option is to try to get a fair share of the economic resources that Westerners possess and produce. With regard to the ECP, Cofán expectations of exploitation were exacerbated by participation in projects that depended on alien activities, relations, and equipment to produce a form of knowledge that was of more use to the outside world than to the community.

No matter how much idealistic practitioners protest to the contrary, the people of Zábalo believe that Western conservationists derive value from Cofán lives and Cofán forests. Obviously, they are right. ECP personnel earn a paycheck for their efforts. They treasure the biodiversity of Cofán territory. And they hope to transform Zábalo experiences into a general model for community-based conservation. From the Cofán perspective, all of the benefits represent the extraction of a surplus from Cofán activity.

ECP staff member planning a trip into Cofán territory (Photo by Michael Cepek)

When pressed, most people in Zábalo admit ignorance as to why outsiders come from so far away to work on conservation. Currently, most Cofán explain their understanding of the word *cientista* (Spanish for "scientist") by extension (i.e., by naming ECP personnel). The most experienced project workers suggest that a scientist is one who "learns everything" about "how things are" and "how things become." Although many ECP personnel were far too routinized in their work to evince touristlike fascination with wildlife, some—such as Debby Moskovitz, whom all project workers recognize as the true ECP *na'su*—exhibited enthusiastic attachment to the *tsampi*. They watched birds, they wandered through the forest alone, and they expressed sincere concern for environmental destruction. In the words of one worker who knew Moskovitz for years, "She really hates the idea of hurting animals. She really hates everything like that. She really loves the *tsampi*. Her true desire is for the *tsampi*. The house of her heart is the *tsampi*. She really loves it. The *tsampi* as well as animals, everything, turtle, woolly monkey, everything. All that is of the *tsampi*."

Even when they recognize outsiders' earnest attachment to their environment, the people of Zábalo cannot imagine any scientist coming to their territory without earning a living by doing so. The Cofán know that scientific conservation is the "work" of ECP personnel. Moreover, they are perpetually aware of Westerners' wealth. The Cofán find it most convincing to use cash as the surest means of calibrating values across deep social divides. No matter how much Cofán people believe that the ECP and others appreciate the *tsampi*, and no matter how much they understand outsiders as idealistic individuals who want to help them, they cannot imagine non-Cofán conservationists doing anything without making money. In the words of one project worker:

> Why would you just do this work if I were the only one being paid? Let's say you look for a project. And then $10,000 comes to you. You receive it, and you give me $500, $500, $500, everything to me. With nothing, would you work? No, you wouldn't. My thought is like this. The museum people don't have other work. Because of that, they want to work with the Cofán, to help the Cofán, and to help other people and other lands. "With that, I, too, will take some of the money." So that they can live. Thinking like that, the museum people want to work here.

This interpretation, more than anything else, explains why Zábalo residents understood ECP-supported projects as work that they performed for the outside world—and for which they deserved a paycheck. Some Cofán are beginning to understand that only *números* can communicate their intact environment and sustainable modes of resource use to an encroaching oil company or a vigilant state environmental ministry. Nevertheless, they have never had to offer such justifications for their land claims. Moreover, their knowledge of the raw power relations and basic unfairness of state machinations leaves them uncertain that they will ever have to—or that it will do any good. And if the portable forms of knowledge produced through scientific conservation are of little direct use to them, they are even more certain that their project participation obeys a logic that is very different from that of their preexisting system of forest care.

In contrast to Zábalo residents, ECP personnel believed that the activities they supported were to be done by the Cofán and for the Cofán. They found it difficult to understand why the people of Zábalo did not embrace scientific knowledge production as a helpful addition to community conservation. Not surprisingly, the ECP staff suffered mounting frustration at what they interpreted as Cofán people's failure to conceptualize the project activities as "theirs."

By the end of 2005, the ECP began to shift its Cofán-centered work from community-based management projects toward large-scale biological inventories of potential conservation areas in threatened Cofán territories. One such inventory convinced the Ecuadorian state to create the Cofán-Bermejo Ecological Reserve in 2002. A 2008 inventory helped the FSC and FEINCE solidify administrative and management rights over the Río Cofanes Territory. Slowly but surely, the ECP learned that its most important contribution was to use its international prestige to produce authoritative reports that can help the Cofán nation acquire more lands as both ancestral territory and protected natural areas. After more than six years of work in Zábalo, the ECP finally understood Cofán perspectives on the utility of science in community-internal contexts. Accordingly, they stopped supporting the projects, and the monitors and coordinators stopped receiving paychecks.

A GLOBAL VISION

Despite their questioning of conservationist intentions and actions, Cofán people see great potential for their relationship with outside institutions. Even though most NGOs understand payment to play a minor role in community conservation, the savviest Cofán leaders hope to construct a much larger collaborative system on the basis of political-economic reciprocity. They realize that their forests matter to Westerners. Furthermore, they know that Cofán people are becoming increasingly adept at producing scientific knowledge, which they want to make available to outside researchers as long as the exchange is balanced.

In short, Cofán activists want to convince the world that Cofán people are the best custodians and investigators of the Amazonian environment. Instead of negotiating short-term interventions aimed at the unrealistic goal of project self-sufficiency, Cofán leaders seek to create permanent partnerships that recognize the reciprocal costs and benefits of scientific research and conservationist practice. In return for protecting and analyzing their forests, they expect steady but modest compensation as well as the political aid that will help them solidify control over their traditional territory.

Cofán leaders hope to take over many of the roles that are currently inhabited by better paid and more securely employed outsiders, whether NGO workers, academic scientists, or state enforcement agents. To date, they have made significant strides toward realizing their vision. With its park guard program, the Cofán nation is directing the management and protection of more than 430,000 hectares of forest. In Zábalo, Cofán project participants received coauthorship recognition for articles in prestigious journals (e.g., Townsend et al. 2005). From the perspective of Cofán leaders, Cofán people's intimate knowledge of and dependence on the forest make them the perfect agents of effective research and conservation. They hope to build on their experience to create new partnerships in Ecuador and abroad.

Although many Zábalo residents were critical of ECP-supported projects, they see great potential for indigenous-environmentalist collaborations. In the eyes of Cofán people, scientific conservation is difficult work. Nevertheless, they are beginning to realize its worth in the wider world. It has provided them with a small but appreciated form of income that allows them to remain in their communities, to avoid harming the forest, and to communicate the fruits of their management practices to a global audience.

Although the ECP ceased supporting Zábalo projects in 2005, some residents who learned scientific techniques began to apply their skills in the park guard program, which involves extensive monitoring activities. The FSC found other institutions to support former ECP projects and related offshoots. The river turtle repopulation effort continues. A group of Zábalo women received training and aid to experiment with new forms of feeding the turtles in the head-starting pools. Some Zábalo residents now work to teach Cofán techniques to other indigenous communities along the Aguarico River and in more distant locations. An Ecuadorian NGO funded a vegetation-mapping project in the community, which employs a number of people. Others became involved in monitoring projects that focus on harpy eagles and giant river otters. In all the collaborations, Cofán people work closely with Western scientists, further mastering the tools and logics that allow them to produce data that is compelling and necessary for global conservation initiatives.

To live their preferred life, Cofán people do not need exorbitant salaries. Nevertheless, they expect compensation for laboring on projects that help the outside world at least as much as they help the Cofán. They hope to maintain their *tsampi* for themselves and for others. To achieve their goals, they need political and economic support. Although they are eternally hopeful that tourists will return, they are preparing for alternatives. From their perspective, becoming scientific conservationists is their best option. With help from the FSC, they believe that their work will continue. As one former coordinator for the ECP projects explained, "Yes, now we're working well. And people from very far, from all over, see our work. And from there, too, they will give money so that we do more. Again, more *semamba* comes to us. It's going well, this *semamba*, this *proyecto*."

6 The School in the City

On a December morning in 2001, I used my satellite telephone to call Quito from Zábalo. Word had arrived that Randy Borman was seriously sick. Amelia Quenamá answered. She told me that her husband was bedridden with a crushing, days-old headache. He had been to multiple Quito doctors. None offered a diagnosis, and his condition was worsening. I handed the phone to Bolivar Lucitante, the community president. After discussing the illness, Lucitante and Quenamá decided that I should bring Cofán plant medicine to Quito, where I would stay before returning to Chicago for Christmas. Zábalo residents collected three bundles of plants. On our canoe ride to Lago Agrio, we stopped in Doreno to collect more medicinal offerings. On the Aguarico, Borman's ill health had become common knowledge.

Before leaving Zábalo, I spoke with Lucitante and Antonio Aguinda[1] about the situation. They were extremely anxious. Although they hoped that the medicine would help, they suggested that Borman's best option was to seek the services of a powerful headwaters shaman. Borman's immobile status in the capital, however, made shamanic intermediation unlikely. Lucitante and Aguinda did not know which enemies were attacking. They assumed that hostile Napo Runa or Secoya *curaga* were behind the assault. In recent years, Zábalo had become embroiled in contentious land conflicts with both peoples.

Considering Borman's fragile state and possible demise, Lucitante and Aguinda were uncertain of Zábalo's future. Without a true *na'su* of either the community or the FSC, they feared an impending disaster. I asked whether Roberto Aguinda would be able to replace Borman. They replied negatively. They said that his imperfect Spanish and lack of other political skills made him an unsuitable candidate. Then, I suggested that the oldest sons of Borman and Lucitante, who were studying in outside high schools, might be able to take over some of the political work. Again, they dismissed the possibility. They explained that the young men had not finished their studies. In short, Borman was still irreplaceable—and still indispensable.

Luckily, Borman did not die. It turned out that a probable case of equine encephalitis caused his pituitary gland to stop working, necessitating a slow recovery with hormone replacement therapy. Gradually, Borman regained some of his strength and assumed his former position. Nevertheless, the incident made people acutely aware of the need to reproduce his skills in other individuals. One Zábalo resident explained, "Truly, we only have one knowledgeable, capable person like Randy here. Because of that, we really want a lot of people, my own children or other people's children, to learn and then to become the supporters, the helpers, of this community, as well as of all Cofán land and the other Cofán communities. With this, we will exist securely." Another man put the situation in the starkest terms, stating simply, "If Randy dies, how will we live?"

In this chapter, I investigate Cofán people's struggle to turn their children into multilingual and multicultural leaders. The effort is mediated by cultural assumptions concerning childhood and socialization as well as the ideological associations of shamanic apprenticeship. Earlier experiences of schooling in Cofán territory, moreover, inform understandings of the urban institutions that represent the best hope for producing Cofán youths with the ability to become global activists. Residing away from their families and communities represents a real risk, however, and no one is sure what the ultimate fruits of the FSC's "Education Project" will be. Nevertheless, all Cofán realize that the formal education of their children is a necessary effort. Only a handful of students have completed the process. Although they are still young, it is possible to learn from their experiences to understand the logic and prospects of the struggle to create a generation of leaders who can replace "the gringo chief."

CHILDHOOD, SOCIALIZATION, AND SCHOOLING IN ZÁBALO

The A'ingae word for child is *du'shu*. The Cofán describe small children as "unripe." From the Cofán perspective, children are paradoxical beings. They display exaggerated symptoms of subjectivity, such as crying, yelling, complaining, and fighting. Nevertheless, people believe that they perform the expressions without reason or cause. Men and women fawn over infants and appreciate their humorous, albeit irrational, behavior. After a child begins to walk, however, people react negatively to its selfish or disruptive ways. Although people sometimes resort to the use of stinging nettles to discipline young children, a more common form of teaching involves reasoning with a child through explicit dialogue on the consequences of improper action. Eventually, a child becomes *injama'pa* (capable of thought and controlled behavior).

People sometimes refer to "difficult" children as *injamambi* (unthinking idiot) or *sumbi* (fool or crazy), two terms that they also use for *curaga*—out of earshot, of course. The similarity between a disobedient child and a powerful shaman is that they are prone to violent, asocial behavior whose negative consequences they do not fear. An improperly socialized child is a raw object that people must make into an *opa a'i*. In contrast, an aspiring shaman seeks to lose his *opa* status by submitting to a process of secondary socialization. He suffers through intense immersion in terrifying realms to emerge on the other side of Cofán sociality.

As soon as they are able to descend from their mother's carrying slings, children spend much of their time in each other's company. They swim in the river and play on the cleared ground between houses. The majority of a child's enculturation, however, occurs through observation and involvement in adult activities. From an early age, children accompany their parents and other relatives during garden work. They watch older people butcher, cook, collect plants, and construct houses. From the age of three or four, boys and girls begin to fish in collective outings. At the age of eight or nine, they accompany older people on hunting trips. By the time they are nine or ten, all girls and many boys can produce the daily staple of banana drink.

A girl leaves *du'shu* status and becomes a *dusunga* (unmarried sexually mature person) after menarche, which occasions the central rite of passage in Cofán society. Upon discovering a girl's first menstruation, people describe her as a *vueyi coen'cho* (recently grown one). The status change typically occurs at the age of twelve or thirteen. Traditionally, a *vueyi coen'cho* is secluded in a ritual hut for the entirety of her first menses. The oldest women told me that the prior custom was to seclude *vueyi coen'cho* for up to four months. The girl's mother and grandmother attend to her, using stinging nettles to hit her legs and form a mat on which she sits. Before the introduction of tampons and pads, a *vueyi coen'cho* would bleed onto her skirt, which she washed in a pot with water from a nearby stream. The girl also bathes with the leaves of *fiño* (*Inga* sp.). While in her hut, a *vueyi coen'cho* begins the dietary prohibitions. The *se'pi'cho* proscribe the consumption of nearly all meat. Cofán people believe that animal blood will cause the girl to bleed excessively and to suffer heavy, frequent, and potentially fatal periods for the rest of her life. In the past, girls followed the prohibitions for two or three months. Today, most only maintain them for one month, if not less.

Upon finishing her first period, a *vueyi coen'cho* washes her body with water boiled with *záttu'cco* (unknown species), the same plant that a postpartum couple uses to purify themselves. The girl gets up from her symbolically seated position. Her caretakers again hit her legs and hands with stinging nettles, "so that she does not become lazy." After the girl has completed the ritual

Cofán girls enjoying a sunny afternoon (Photo by Michael Cepek)

treatments, an especially hardworking older woman instructs her in typical female responsibilities. They wash clothes, cut firewood, and clear vegetation around the house. The publicly performed acts symbolically mark the *vueyi coen'cho* as an adult who can take a husband and sustain a household. At that point, people say that she has *ccasheve da* (become an adult woman).

A boy's entrance into *dusunga* status is less marked. The only physical transformation that correlates with menarche is the deepening of a boy's voice, which occurs between the ages of twelve and sixteen. The verb that refers to the change is *shagattuye*, which is related to the term for tropical cedar tree, *shaga'to*. Some boys perform a solitary ritual when they *shagattu*. With no one watching, they dive into the river and emerge at the tail end of a canoe, many of which are constructed from *shaga'to*. Then, they pull themselves above the water and scream loudly, using no words. They perform the ritual to "close up" their throats. If a boy does not complete the action, people say that his throat will become excessively large. Consequently, his voice will be too deep and too loud. As with the menarche ritual, the logic is to prevent a bodily transformation from going too far, which would produce a problematic relation between internal and external spaces, as mediated by the key passageways of the vagina and the mouth.

There is no ritual instruction of a *dusunga* boy. Nevertheless, when he *shagattu*, people expect him to start looking for potential romantic partners. Typically, a father counsels a *shagattu* boy, saying, "Look, what kind of a *du'shu* are you? You're still a *du'shu*. If you get married, how will you care for your wife? You still don't have a garden, you still don't know how to fish, and you still don't know how to hunt. Why do you think like that? Just let it be." A *dusunga* boy begins to hunt frequently during the period, and he also learns traditional male activities, such as canoe construction. *Dusunga* boys feel that young women are watching them. They believe that if girls see them coming back from the forest with a lot of game, they will think, "I really want to marry that person. If I marry him, we'll have a lot of meat, and I'll care for him well and we'll live that way." In addition, some *dusunga* boys share meat with potential parents-in-law, who might favor a marriage because of the informal gifts.

Marriage constitutes a person as a *coenza* (adult or old person).[2] Normally, a man resides with his new father-in-law for a short period and performs a minimal bride service, such as aiding in the construction of a house or a canoe. The bride service does not always occur, however, and both families engage in long discussions to decide where the newlyweds will live. Most people agree that the decision is based on two considerations: first, the desire of the young couple themselves; and second, the composition of the respective families. If one set of parents has no sons, their daughter and son-in-law will move in with them. The newly resident man will help his father-in-law by contributing meat to the household and working alongside him on daily tasks. If one set of parents has no daughters, their son and daughter-in-law will move in with them. The newly resident woman will help her mother-in-law to clean and prepare food. After living in their parents' home for as long as a few years, the young couple build their own house. Nevertheless, they typically remain in the same *naccu*, gradually developing their own gardens over a period of five to ten years. When they have their first child, they receive teknonyms as their primary terms of address and reference, becoming, for example, Ignacia *mama* (Mother of Ignacia) or Ignacia *yaya* (Father of Ignacia). With new names, they are full adults.

Shamanic Training

Du'shu and *dusunga* are free to stay in their homes or to move between households, *naccu*, and communities. Traditionally, shamanic training was the only activity that necessitated long-term absence from one's natal family. Currently, few Cofán youths want to become *curaga*. Nevertheless, Cofán continue to imagine it as a possibility for young people who take an interest in *yaje* and show courage and discipline. Collective drinking provides children as young as seven with the opportunity to demonstrate their desire to embark on the shamanic path. When I drank *yaje* one night in Zábalo, the officiating elder offered a gourdful to his grandson, who was nine. The boy became afraid. Nevertheless, his older brother, who was home on vacation from a Quito school, decided to drink.

If a young person can drink without fear, their father or grandfather might take them to be taught by a powerful *curaga* in another village, whether populated by Cofán or other peoples. The most renowned Cofán shamans trained with Siona and Secoya masters. They learned alien languages and customs while residing with distant *curaga* for periods of more than a year. Shamans who instruct outside youths enhance their reputation. Moreover, the apprentices help with the menial tasks of boiling *yaje* and sustaining a household. Furthermore, a father might "pay" for his child's shamanic education with a feather headdress, a prized possession that represents hundreds of hours of labor.

Becoming a *curaga* is the prototypical form of learning in Cofán society. The lexeme for both "to learn" and "to know" is *atesuye*, and the main A'ingae designator for a powerful shaman is *atesu'cho* (one who knows or has learned). Only a few people successfully complete the instruction. The process requires and produces their atypicality as fearless individuals who identify more with a jaguar than an *opa* woolly monkey. For the neophyte, the *curaga* becomes a "teacher," a "producer of visions," and a "declarer of prohibitions." The marked asymmetry of the shaman-student relation is unique in Cofán society. Its extremity is more pronounced than the power of a parent over his or her infant child.

Under the guidance of a *curaga*, a student lives a life of intense privation. In addition to residing away from family and friends, he eats few types of food—and in extremely limited quantities. He has no romantic contact with members of the opposite sex. He must cautiously avoid contaminating spaces, objects, and individuals. As instruction proceeds, the *curaga* and the *yaje a'i* command the student to drink more and more hallucinogens and emetics. They urge him to confront and transcend his fears of death. If a student does not succeed at becoming a *curaga*, people invariably find the explanation in his inability to live a life of hunger, isolation, discipline, and terror.

Shamanic instruction provides the referent for the key concept of Cofán pedagogy: *vanamba atesuye* (to learn by suffering). The idea of learning by suffering guides Cofán people's understanding of the processes to which their children must submit if they hope to master the knowledge associated with external domains. Like shamanic instruction, school knowledge involves other languages, other peoples, and extra-community learning. Unwillingness to *vanamba atesuye* explains many Cofán children's disinterest in entering an outside school or finishing their formal education, as well as their desire to avoid the shamanic path.

During my fieldwork, one Zábalo father tried to convince his teenage son to persevere with his education at a high school in an upriver Secoya community. The *dusunga* felt hungry and isolated in the house of a distantly related Cofán man who had married a Secoya woman. His father explained to him: "*Estudio* (school knowledge) is something that is only learned through suffering. Truly, you will not know the satiated life that you knew in your home. That's the way one suffers. You must suffer and learn." After little more than a year, the youth dropped out, got married, and returned to an *opatssi* life in his father's *naccu*.

Reflecting on the difficulties of Cofán children who study in Quito, Borman explained, "It's understood that this is a sacrifice time of your life. It's understood that this is the same sort of sacrifice that's being asked of a shaman's apprentice. We're talking about dedication. We're talking about abstention. We're talking about a number of things that are already present in the culture, under other circumstances."

In short, for Cofán people, to learn is to suffer. It demands departing from the company and care of one's primary household and extended family. Furthermore, it entails a process of transformation that results in moral and ethnic ambiguity. People feel ambivalent about the children they send to either a distant *curaga* or an outside school. On the one hand, people have deep attachments to the children. When they cannot assure their care, they worry about their health and happiness, and they know that the children are suffering. On the other hand, Cofán people have always been open to the possibility that their children might leave their homes temporarily or permanently. Moreover, they accept that a transformed child might not work to ensure the well-being of his family or home community. Nevertheless, Cofán people realize that they must take the risk if they hope to produce individuals who can negotiate the threatening difference of a hostile world.

Schooling in Zábalo

Cofán people know little about the history of schooling in their territory. They often refer to the tales of Guillermo Quenamá, Doreno's founding shaman-chief. As a child at the turn of the century, he was brought by Catholic priests to a school on the San Miguel River. Teachers and students ridiculed him for being a "monkey," and they punished him in humiliating ways. He did not stay for long. He was very hesitant to let Bub and Bobbie Borman create an A'ingae-language school in Doreno. Eventually, he relented. At first, the

Summer Institute of Linguistics funded the school, which taught basic literacy and numeracy. By the late 1960s, two Cofán teachers, whom the Bormans had trained, began teaching. By the late 1970s, the state took over education in indigenous communities. In 1980, it approved bilingual education and put the Provincial Center of Bilingual Education in charge of Cofán schooling.

Cofán schools exhibit a number of problems: Teachers possess a low level of education and training; schools suffer a severe shortage of resources; and few Cofán teachers feel proficient in Spanish, the acquisition of which the Cofán view as a key goal of formal education. For all these reasons, some Cofán parents want to send their children to schools in non-Cofán communities, whether indigenous or Cocama. Nevertheless, few families can afford the expense, and even fewer students can deal with the distance. As a result, even in 2012, only a couple dozen Cofán people have a high school degree, and only Borman's eldest son has finished college.

Zábalo's primary school was created in the late 1980s. People refer to it as the "learning thing." It offers grades one through six. Usually, there is only one teacher. Despite its small size, the school is the most salient sign of state presence in Zábalo. It contains books, posters, chairs, desks, and blackboards that were provided by the Ministry of Education. Moreover, the federal government distributes beans, rice, tuna, oatmeal, and sugar to students. As in other areas of Amazonian Ecuador, the state hopes that its support of school instruction will transform its peripheral peoples into modern citizens (Rival 1996).

As with all Cofán schools, Zábalo's school suffers a lack of materials. Many children do not have sufficient pens, pencils, and notebooks to complete their work. For families who have no income and five or more children, even the minimal annual cost of $20 per student is too much.[3] The biggest problem with Zábalo's school, however, is the unavailability of good instructors. Few Cofán teachers have a true high school education. Many speak little or no Spanish. And there is virtually no state control over teacher certification. One Zábalo teacher told me that the only requirement is to take a test of A'ingae proficiency in Lago Agrio. Although many Cofán can pass the test, few are willing to work as teachers. The job represents long hours, an inflexible structure, and low pay. Until a regional strike in 2001, beginning teachers earned $40 a month.

The most deep-seated problem with Zábalo's school, however, is its complete immersion in the community context. As an institution, it does not foster intensely disciplined practice. Its spatial, temporal, and social boundaries are fuzzy. The school day is supposed to begin at 8 AM and end at 12:30 PM. Nevertheless, teachers and students come late and leave early. On some days, teachers do not arrive at all, as they have to hunt or garden in order to produce food. Furthermore, teachers sometimes attend training courses in Lago Agrio. They can be absent for weeks at a time. Given difficulties in travel between Doreno and Zábalo, no one is sure when the teachers will return. Departure dates, too, are uncertain. Most teachers have to join any canoe trips they can, given their lack of a boat, motor, and gasoline of their own.

Aside from the temporal concerns, Zábalo's school is socially and spatially open to the rest of the community. Adults and nonstudent children walk in and out of the school to pass time or to talk to friends and relatives. Many students leave the school to run back to their homes, often without explanation. The students' households, moreover, are poor environments for completing homework. The majority of parents did not finish school, and they supply little guidance and discipline for their children. Apart from providing scolding, counseling, and punishment in response to violent or disruptive behavior, Cofán parents are not harsh disciplinarians.

Teachers, too, are poor disciplinarians. Unlike a *curaga* who can inspire fear in his apprentice, teachers are usually too young and too *opatssi*. Parents tell them to scold and even physically punish students. Most of them, however, are uncomfortable with the practice, which the Ministry of Education forbids. One older parent gathered stinging nettles and gave them to a young teacher, who laid them on his desk without mentioning the possibility of punishment. The teacher reported that just looking at the nettles improved student behavior, at least temporarily. With one teacher handling children that range in age from seven to sixteen, the classroom is difficult to manage. The situation requires substantial self-motivation. Students must work alone on tasks while the teacher instructs children of different levels. Unfortunately, many students lack the required discipline. They progress slowly, if at all.

Most Cofán feel ambivalent about bilingual education. Many understand that children more easily learn the basics of literacy and mathematics in their own language. Nevertheless, everyone points to the obvious fault of Cofán schooling: not one student has learned Spanish in a Cofán school. Mastering the language continues to be a central goal of formal education. In addition, bilingual education presents an additional problem, as it seeks to base lessons on "culturally suitable" topics, such as mythology, subsistence, and Cofán custom. Consequently, it contributes to the central challenge of Cofán schooling: its complete immersion in a Cofán lifeworld.

Relative to shamanic training, schooling in Cofán territory is inherently problematic. In schools that are subsumed by *opatssi* community life rather than opposed to it, to learn is not to suffer. Students do not reside away from their friends and family, and they do not enter a process that results in moral and ethnic ambiguity. Given that the acquisition of other-language and other-knowledge is the desired outcome of school-based learning, educating children in Cofán communities—with "Cofán" themes and materials —is the opposite of traditional Cofán instruction, which is based on the template of shamanic apprenticeship. At some point in the future, the quality of teachers and materials might be sufficient to make community schools suitable learning environments. For the time being, however, the institutions are incapable of producing students who can meet the demands of contemporary political practice.

QUITO SCHOOLS

The predicament of Cofán schooling revolves around the contradiction between living in a Cofán community and acquiring the knowledge and power of ethnic others. Schools in Zábalo and other Cofán communities do not function effectively because they are insufficiently separated from a Cofán lifeworld. Quito schools, in contrast, are extremely effective at transmitting outside knowledge and abilities, but they reflect no aspects of Cofán culture. Consequently, they pose the twin risks of excessive culture change and unbearable culture shock. The production of contemporary *na'su* depends on a successful negotiation of the double danger.

The first Cofán students to enter Quito schools were Borman's three children, two of whom entered the Alliance Academy in 1998. At the same time, four Cofán students entered the Fundación Interayuda, an Ecuadorian-run Christian boys' home and school in the nearby town of Tumbaco. A Quito man who knew the people of Zábalo from his work with the Flotel Orellana introduced the community to the institution. Apart from transportation costs, the Tumbaco school is free. Borman decided that entering Cofán children in the program was a worthwhile experiment. He and Quenamá live close by, and they brought the students to the QCC on weekends, where they socialized with other Cofán.

After a year in Tumbaco, three of the first students moved into the QCC and entered the private Escuela del Futuro. At that point, the FSC's Education Project truly began. Although the project has received little outside support, it is one of the most important Cofán efforts. Its central goal is to place Cofán children in a number of educational institutions. The first phase of the effort involved Borman's children and a small core of students who attended the Escuela del Futuro and then the Colegio Francés, another private Quito school. My main fieldwork years of 2001 and 2002 overlapped with this period.

The two most important institutions of the Education Project's early years were the Alliance Academy and the Colegio Francés. Although the Tumbaco school continues to receive students, its quality of instruction is very low. During my time in the field, it served as a "testing ground" to identify capable children, the best of whom entered the Colegio Francés. A few years after I left the field, Cofán students stopped attending the Colegio Francés. Borman, Espinosa, and López identified a more affordable school that was closer to the QCC, which allowed students to transport themselves with their feet rather than a bus or a taxi. I did no research at the latter school, which Cofán students are still attending. In addition, Cofán students continue to study at the Alliance Academy, the highest-quality and most expensive institution in the project. Only the best Cofán students make it to the Alliance Academy, which prepares them for a university education in Ecuador or abroad.

Colegio Francés

The Colegio Francés sits at the far northwest of Quito in the town of Pomasqui, a short bus ride from the QCC. It offers instruction from grades one through twelve. During 2001 and 2002, tuition was $75/month. Nevertheless, Borman estimates that extra expenses (e.g., books, materials, uniforms, etc.) raised the actual annual cost to $1,500 per student. The FSC covered the great majority of the expenses. At the time, Zábalo families paid the FSC $30 per month for each of their participating children. The FSC covered remaining education costs as well as the students' living expenses with its general operating funds, which are always in short supply.[4]

The Colegio Francés was founded by a group of Ecuadorian students who studied in France in the 1950s and 1960s. They returned to Ecuador to create a school with a bilingual French-Spanish curriculum. Currently, only the name exists as a reminder of the original mission. The school now places primary emphasis on English. In general, the Colegio Francés serves middle-class *quiteños*. Its teachers and administrators try to distinguish it by embracing an environmental mission. Its official slogan is "An education for life in harmony with nature." The school devotes substantial resources to maintaining its lush grounds, and students are prohibited from breaking a twig or leaf while on campus. There is an active sports program, and the school has good facilities as well as mandatory computer instruction.

Although I am not trained to judge teacher performance, the instructors whom I watched while shadowing Cofán students appeared to be well educated, conscientious, and successful at maintaining a disciplined learning environment. Only the English instructor—a native Ecuadorian—seemed less than qualified for her subject. The Cofán students took courses in English, social studies, mathematics, computers, natural science, physical education, and language (i.e., Spanish vocabulary, grammar, and rhetoric). The Colegio

Francés seeks to put its students, Cofán or otherwise, on the same educational level as Ecuador's urban middle class.

On the surface, it was hard to distinguish the Cofán students from their classmates. Teachers and administrators stated that the Cofán youths acted "normally," "like other children." Some teachers noticed small problems with their Spanish and verbal reasoning, but they claimed that their abilities were within the normal range. The Cofán students themselves asserted that language difficulties made their work excessively difficult. Generally, however, they did fairly well. Only one had to study over the summer to avoid repeating a year. The performance of another was among the best in his class. As a whole, the Cofán were "B" or "C" students.

All teachers told me that they were proud to instruct Cofán students. Echoing a common perspective, the rector said, "I believe that it is an honor to have Cofán children in an institution where we can rescue a little bit of the richness of our Ecuadorian culture." The Cofán students reported few cases of being taunted or criticized by teachers or classmates. While observing them inside and outside of the classroom, I noticed no overt prejudice. If anything, students, teachers, and administrators wanted to hear them speak more about their *cultura* and *identidad*. The Cofán youths, however, were hesitant to discuss the topic. In interviews, none of them said that they were ashamed of being Cofán. Instead, they claimed that they were too lazy to explain their background to Cocama. More importantly, they did not enjoy becoming objects of excessive attention.

Multiple teachers told me that the Cofán students stood out in one respect: their stellar performance in sports. People admired them for their speed and strength.[5] During my classroom observations, Cocama students appeared to like the Cofán. I caused the only problematic interaction that I witnessed by teaching Cocama students the A'ingae word for "hello." They responded by laughingly shouting the word at the Cofán whenever they walked by. The interaction made the Cofán youths feel awkward. They shrank from the attention, even though there was nothing negative or prejudicial about it.

In general, the Colegio Francés provides a much better education than schools in Cofán communities. Nevertheless, neither FSC employees nor other Cofán people know what to expect from a Colegio Francés–type education. They know that students will not be able to speak English. They assume, however, that Spanish reading and writing skills will be relatively good. In addition, they expect the students to have a heightened facility with mathematics and computers. Moreover, Cofán people assume that attending middle-class mestizo schools will familiarize the students with the ways of city Cocama, who inhabit the bureaucratic networks that Cofán leaders negotiate.

Although the Cofán students were relatively at ease at the Colegio Francés, they did not establish lasting social relationships with Cocama classmates. Much of their Quito experience is marked by a feeling of isolation. Their most striking form of discomfort arose after manifesting an overt sign of Cofán-ness (i.e., use of A'ingae). They did not want their ethnic difference to become the topic of Cocama commentary, even if friendly in nature. Their discomfort was a sign of their lasting *opa*-ness, which they must transcend if they are to practice a form of politics that depends on the recognition of their ethnic difference in local, national, and global arenas.

The Alliance Academy

Unlike the Colegio Francés, life at the Alliance Academy bears an oblique relation to the cultural practices of middle-class, urban Ecuador. A small group of North American evangelical missionaries formed the school in 1929. Their main objective was to provide an education that would enable missionary children to enter colleges in the United States. They set out to create a high-quality Christian school based on a North American curriculum. The Alliance Academy offers instruction from kindergarten through twelfth grade. In 2002, its student body exhibited an impressive ethnonational diversity: 50 percent from the United States, 32 percent from Ecuador, and 18 percent from Brazil, Canada, China, Colombia, El Salvador, Japan, Korea, Norway, Paraguay, Peru, Sweden, Taiwan, Trinidad and Tobago, the United Kingdom, and Venezuela.

Initially, only missionary children enrolled at the Alliance Academy. By 2002, however, 40 percent of its students had no missionary affiliation—and none need be Christian. Many students attend the school because of its English-language setting, its good academic reputation, and its strong ESL program. A large number come from families associated with multinational corporations and the diplomatic community. The Alliance Academy offers a much more affordable education than Quito's other top English-language schools, making it an even more attractive institution.

Borman's children and other Cofán students are eligible for a 50 percent "missionary discount." In 2001 and 2002, they paid $3,200 per year. The high cost restricts the number of Cofán who can attend. Moreover, the language requirement is an obstacle. If a child enters at grade six or higher, he or she must have a third-grade reading level in English. Nearly all Cofán children cannot pass the test, including the students who started at the Colegio Francés. There is some leeway with the requirement, however. Any student who enters at an earlier age or who enters at the desired level can enroll in the ESL program. The specialized instruction provides a helpful class structure, intensive language instruction, qualified grading, and in-class tutoring for three years.

Although the high cost represents a real challenge, the Alliance Academy offers many benefits. Its facilities and resources are much better than what the Colegio Francés offers. College-educated North Americans, who are fluent in English, teach the majority of classes. The school uses North American textbooks and materials. It possesses a large library, and class sizes are small. The school is equipped with the latest educational technology. It has multiple labs stocked with Apple computers. Students do not have to share units or take turns, as they do at the Colegio Francés.

The Alliance Academy's student body is also much more diverse. Students do not wear uniforms, and there is an immense amount of ethnoracial heterogeneity. As I accompanied Cofán students through their daily rounds, I heard numerous languages (e.g., Spanish, English, Korean). Nevertheless, the Alliance Academy's Christian character ensures a strong element of homogeneity. Students attend chapel once a week. There is Bible class every day. And spiritual messages cover the walls. The teachers and administrators to whom I spoke, however, stated that the Alliance Academy does not "force" beliefs on anyone. If they have the confidence, students are free to argue their positions in Bible class and refrain from prayer.

Despite their stated tolerance, the great majority of teachers and administrators consider their jobs to be their "missions." Some are affiliated with the Christian and Missionary Alliance. Others are supported by home congregations. Most of the coursework, however, revolves around secular materials. Based on my observations and interviews, Alliance Academy teachers have a much more nuanced sense of cultural diversity than Colegio Francés instructors. In Felipe Borman's ninth-grade English class, the students did a section on missionary history. They watched *The Mission* and read a Japanese novel about Jesuit boarding schools. In addition, they read one of Amy Tan's books during a section on multiculturalism.

Although many Alliance Academy teachers attended Bible colleges, they spend much of their lives in diverse contexts. They are extremely wary of "the great white father" stereotype. In one teacher's words, "Missionaries are usually far away. And so if the average North American actually meets one, what have they drawn with their schema? Usually it's Hollywood. And Hollywood is working on something that disappeared a hundred years ago." Whereas Colegio Francés teachers and administrators embrace a folkloric sense of culture (i.e., celebrating traditional dress and dance), Alliance Academy teachers exhibit a more complex understanding. When I asked one teacher whether Cofán children would undergo "culture loss" by attending the Alliance Academy, he replied:

People who think that have a romantic idea about cultures. I don't think that any cultures are static, unless they're isolated. As long as you have two cultures that are commercially interchanging or whatever, you're gonna have growth, you're gonna have exchange of ideas. I think for us, the idea of having a native tribe that looks like something out of the eighteenth century is beautiful and romantic. I don't think it really exists. I think that cultures were always changing. And today, if people are worried about them losing their culture because they'll be educated and exchange ideas with the First World—yes, that might happen. It probably will happen to a certain degree. But if you do not prepare them to interact with First World cultures, I could practically guarantee that they will be lost. They will lose all their cultural identity. They will just become a third-rate gofer with some oil company, or something like that. Yeah, I think that people feel that we shouldn't do it. But they're not going to do anything for the Cofán anyway. Except feel that they've done the right thing of letting a romantic idea persist that doesn't even exist.

Like the Colegio Francés teachers, Alliance Academy faculty feel that it is an "honor" to instruct Cofán students. Partially because of their intercultural background and partially because of their familiarity with the work of the Borman family, the teachers know a great deal about Cofán culture and politics. One teacher spent an intriguing week in Zábalo: "It was super. I tell you, I dream about it. It's fascinating. If I could spend a year living down in Zábalo, I would." My sample of Cofán students was too small to hypothesize about the effects of this and other teachers' sentiments. Nevertheless, it was clear that students at the Alliance Academy differed from youths at the Colegio Francés in an important way: they constantly talked about being Cofán.

While I followed Borman's children through their classes, they spoke A'ingae to me and translated their words to classmates. The teachers encouraged the Borman children and Hugo Lucitante (who studied at the Alliance Academy for one year) to include their "culture" in discussions and assignments. The students showed me their stories and poems. They often wrote about fishing and hunting. When teachers asked students to bring foods from their "ethnic backgrounds," Felipe Borman arrived with the less-than-fresh head of a Zábalo fish. He laughed as he ate it in front of his entire class. According to Cofán students, their Alliance Academy classmates think that being Cofán is "cool." Many of their friends express a desire to visit Zábalo. When interviewing Hugo Lucitante's father about his son's experience at the Alliance Academy, he agreed that Gringo really like *a'i*, as he learned through tourist interactions:

Gringo, well, more than any other people, they really like *a'i*. Gringo won't say, "You Indians, you're dirty." They won't speak like that. Gringo come here and they have compassion for *a'i*. They really like us. I don't know why. I've thought about it, "Why do Gringo really like *a'i*?" That's the way they are. When tourists come, they see us, and they're happy. That's good. When Cocama come, they have no compassion. Cocama will insult us, saying, "You Indians live in filth." They'll say all kinds of things.

There are too many variables to draw definitive conclusions about the effects of an Alliance Academy education on Cofán confidence. Borman's children and Hugo Lucitante are special cases. They are fluent in English, and they have spent significant time in multicultural settings. Nevertheless, the Alliance Academy clearly provides a strong context for the establishment of a positive identity in relation to such social others as Gringo, upper-class Cocama, and other foreigners. Any Cofán leader must feel comfortable negotiating relations with individuals from all the groups.

Most importantly, the Alliance Academy offers an education that is on par with upper-tier public and private schools in the United States.[6] Reflecting on Felipe Borman's progress since entering the school as a young Cofán with muddled English, one teacher said, "He'll be able to do what he wants to do, and we'll prepare him for that." After all, the Alliance Academy exists to allay the fears of college-educated missionaries, who do not want to sell their children short on opportunities for higher learning.

Not every Cofán child will attend the Alliance Academy. The school is only willing to extend its missionary discount to six or seven Cofán students. In 2001 and 2002, the FSC did not even have enough funds to cover students at the Colegio Francés.[7] Furthermore, language poses a serious challenge. Unless students can learn enough English in schools such as the Colegio Francés, they will never be able to enter the Alliance Academy, except if they begin at a very early age. Such an option, however, is problematic. Borman and other Cofán feel that it is unwise to invest too much money in the education of children who have not proved their interest, discipline, and aptitude in a cheaper institution. The ideal situation would be to have especially interested children begin in Tumbaco, progress to schools such as the Colegio Francés, and finish at the Alliance Academy, which would mark them as extremely promising figures.

Although Cofán people do not express concern about the Christian nature of the Alliance Academy, the majority of secular outsiders are leery of the institution. Consequently, securing funds to support the education of Cofán children in the school represents a real challenge. Continuing to place many

Cofán students in urban mestizo schools is not without benefits, however. Although Alliance Academy graduates will undoubtedly speak the best English and be the most prepared for entering universities, their Spanish will not be as good as that of students who graduate from *quiteño* institutions. Furthermore, graduates of such schools as the Colegio Francés will be better equipped with the contacts and social skills that enable effective bureaucratic action in Quito. In general, the ultimate point of the Education Project is not to produce one Borman-like *na'su*, but to prepare a group of leader figures with diverse abilities and dispositions, all of which will be important for the success of the Cofán experiment.

LIFE IN QUITO

When Borman created the FSC and the QCC in 1998, he wanted to build a Cofán base in the capital. The QCC became essential for conducting political work, hosting Cofán visitors seeking medical treatment, and providing a home for Cofán students. The center's population can swell to as many as one hundred individuals for short periods. In 2001 and 2002, however, its permanent residents numbered nine: Borman, Quenamá, their three children, and four other Cofán youths.[8] In addition, Roberto Aguinda and his wife, Linda Ortiz, stayed at the QCC for long periods to do political work and to care for their child, who was attending the Colegio Francés. Borman's family lived in the upper half of one QCC building. The rest of the Cofán students and visitors stayed in another building. The complex is located at the northern end of the city in a semi-industrial neighborhood named Carcelén Alto.

Cofán individuals traveled to Quito throughout their history to trade lowland products for manufactured commodities and highland goods. The oldest living Cofán remember the routes that their parents took by foot. Many of my middle-aged consultants rode the bus to the city after Texaco-Gulf constructed the first highway. The creation of the QCC, however, represented a significant expansion of Cofán possibilities. Families and individuals have never had to confront the pleasures and pains of the urban Andean environment on a long-term basis.

Although Cofán students relate to life in Quito as a struggle not entirely unlike the tribulations of shamanic apprenticeship, the city holds its own ambivalent promises. The Cofán, however, do not experience the alien world in an unmediated form. QCC inhabitants reproduce many aspects of a community-based way of life in the urban environment, thereby creating a hybrid space of change and continuity. The mixed lifeworld forms the context in which Cofán youths evaluate their ability and desire to live away from home communities as students and potential leaders.

In 2001 and 2002, the average weekday of Cofán students proceeded along the following lines: wake up at 6 or 6:30 AM to get dressed; eat a breakfast prepared by Borman; take a bus to school at 7 or 7:30 AM; stay at school until 2 or 3 PM; return to the QCC to rest, play, and do homework; eat a collective dinner prepared by Quenamá at 5 or 6 PM; work on school assignments and watch television until 8 or 9 PM; bathe and go to sleep by 10 PM. Weekends were relatively unstructured. On some Saturdays, the whole group left the city to wander through the upper reaches of the Cayambe-Coca Ecological Reserve or to camp close to the Reventador volcano on the Coca River. The group also made day trips to the houses of Randy's brothers Ric and Ron, who live outside of Quito in recreational camps for Christian youth. On many Sundays, the Borman family attended services at the English Fellowship Church. Sometimes, the other Cofán students accompanied them.

I spent my first six months of fieldwork at the QCC. During my stay, I followed the Borman family and Cofán students through their daily routines. I lived in the back building, where I was in close quarters with visitors and the Cofán youth. Although Quito was radically different from the Chicago winter I had just left, I did not find it hard to acclimatize to the new setting. Nevertheless, upon my return to Quito after the second half of my first year of fieldwork, during which I lived in Zábalo, I was overwhelmed by the material, practical, and social peculiarities of the urban environment.

My first night back in the city, I sat at my computer to record the differences: Quito's constant illumination and vague distinctions between day and night; the cold, dry air of Andean Ecuador, which offers instantaneous evaporation instead of rotting moistness; the dense, still background of Zábalo's forest, as opposed to the rush of cars and trucks that clutter the urban landscape; the *tsampi*'s unending hum of rivers, birds, insects, and frogs, as against Quito's parade of crashing traffic and descending airplanes; and the overwhelming sea of anonymous humanity in the city, which felt so different from Zábalo's small circle of named and known individuals.

Many of the children and adults who stayed in Quito made similar comments about the urban environment. Every Cofán person to whom I spoke said that they did not like to be in the city for more than a week. With one and a half million inhabitants, Quito sprawls over an Andean valley at more than 9,000 feet above sea level. Although it feels springlike to me, Cofán people universally state that it is too cold. They find the air too thin for easy breathing. Many say that Quito makes them sick with headaches, earaches, colds, nausea, digestive problems, mouth sores, and dry and bleeding lips. They complain that the city reeks of gasoline and exhaust, that it is too noisy and too crowded, and that it is bare of vegetation and covered with hard concrete and broken glass. One of Borman's sons, who had been living in Quito for three years, shared his opinion:

> In Zábalo, you can run faster without having to breathe so much. And there's more jungle, so you can climb trees and you can plant stuff and you usually are active all day, because you'll be going swimming or doing something at least. Here, right now where we're living, the airplanes, there's a lot. And I don't like them. And there's cars, usually noisy. In Zábalo, there's other noises. And some birds just make beautiful noises, and it helps you to go to sleep, at least to me because I like music or something like that to go to sleep with. Up here there's not really good music, I mean, stuff that's making good music, kind of natural music [i.e., animal noises]. But in Zábalo you know you're in the deep jungle. The air's real. You're warm there. Here, you'll be cold. It smells worse than in Zábalo, real gassy or something like that. And it makes me want to throw up.

Most Cofán feel nervous in the city. While walking in the forest on hunting and gathering trips, Zábalo residents often told me that the *tsampi* is "our city." Many Cofán have gotten lost in Quito. They realize that their experience of the urban environment differs from the mastery of Gringo and Cocama, who are, in turn, helpless in the forest. Cofán people suggest that there is little that can hurt or kill one in the *tsampi*. In contrast, Quito is filled with "thieves" and "killers." Newcomers and long-term QCC residents express their dislike for Quito in terms of their inability to wander in an *opatssi* fashion. They fear becoming lost, being mugged or murdered, or getting hit by a car. One of the Cofán students explained, "We can't really reside in an *opatssi* fashion here in Quito. One will always be afraid." Although the Cofán recognize the dangers of forest life, they realize that their extensive knowledge ensures their safety. As one Zábalo resident said of life in the forest, "there is nothing from which you must die." Quito's human antagonists, however, are virtually impossible to fight off.

Cofán students also find life at the QCC to be exceedingly boring. They experience the QCC as a restriction, rather than an expansion, of practical movement and social contact. In Zábalo, they are free to visit other households, to talk with friends and family, and to make what they want of their days. In Quito, their discomfort with the urban landscape, as well as their lack of money, leads to a sedentary and insular existence. In the words of one student, life in the city involves too much "sitting." Although Borman and Aguinda interact with many people as part of their work, other QCC inhabitants interact mainly with each other.

In short, the significant others of all QCC residents are other QCC residents. No *Quito'su a'i* (Quito Cofán) maintain friendships with Cocama or Gringo that are as important as their relationships with kin and friends in lowland communities. When I asked one Cofán student why *opatssi* life was impossible in Quito, he referred to the social and practical restrictions:

> In Zábalo, you live according to your desires. You have no work there. You sleep, wake up, and think about what you want to do. You can swim for as long as you want, hunt for as long as you want, and you do it. But here, you just sit and stay. You become very sad, and you want to go back to Zábalo. To swim, play, and talk with your friends—that's what you want.

The QCC affords few of Zábalo's pleasures. In the interviews that I conducted with Cofán students in both Zábalo and Quito, I asked them if there was anything that they missed about the city while at home on vacation. I expected them to talk about school friends, television, and the hustle and bustle of the urban environment. Instead, their answers focused on the Quito foods (e.g., pizza and roasted chicken) that are unavailable in the *tsampi*. Nevertheless, they said that they would forget about the desired foods after living in Zábalo for a month. In addition, they reminded me that the items are available in Lago Agrio, to which they can travel while visiting friends and family in Doreno.

Life at the QCC is difficult and alien. Some Cofán customs, however, mediate city life, making it more bearable. Borman and other Cofán leaders want the QCC to be a middle ground between Cofán communities and a Cocama city. They believe that the QCC's structure as a hybrid space makes it easier for Cofán students to deal with the problematic outcomes of excessive culture change and unbearable culture shock. Both would threaten their potential as future leaders.

Most casual visitors to the QCC cannot differentiate the material layout of the space from other Quito homes. Nevertheless, a closer look at the kitchen, for example, uncovers numerous traces of an Amazonian lifeway. On the stove, there is usually a large aluminum pot filled with banana drink. The drawer below the counter contains a number of Cofán implements: a stick for mixing and stirring banana drink; a plantain-grating tool made out of a spiny palm root; and fire fanners made with vulture, curassow, and guan feathers. There is a six-inch stool in front of the sink, as many Cofán are too short to use the countertop without supplementing their height.

Cofán student preparing to fry a caiman leg at the QCC
(Photo by Michael Cepek)

Often, large stalks of Zábalo or Doreno plantains sit on the floor. Many other vegetables and fruits—such as lemons, avocados, peach palm, aguaje palm, manioc, and "wild grapes" (*Pourouma aspera*)—also make the trip from lowland communities. Sometimes, a land tortoise is tied to the cabinet, where it waits to be killed, butchered, and eaten. The freezer is often filled with meat and fish that residents amassed during trips to Zábalo. During the river turtle egg-laying season, bags of the fatty morsels fill the refrigerator. Borman and Quenamá obtain the great majority of the food with their own labor. Other Cofán whose children reside at the QCC periodically contribute meat, plantains, and manioc.

Cofán also bring baby forest animals to the QCC as pets. In 2001 and 2002, woolly monkeys, river turtles, and smooth-fronted caimans lived at the center. Sometimes, the animals appeared in the students' science projects. In addition, there is always a large group of dogs that derive from Cocama and Cofán stocks. Often, one can hear Quenamá or Borman shout the animals' names (e.g., Shipicco [Cockroach] or Sararo [Giant River Otter]) when they bark too loudly. Finally, Quenamá and others spend many afternoons lying in the hammock outside of Borman's upstairs bedroom. They make necklaces and bracelets with palm fiber and the seeds of forest plants, as well as glass beads and fishing line.

All of these objects support activity structures that combine elements of the *tsampi* and the city in a transformed version of what the philosopher Martin Heidegger would call an "equipmental whole" (1996). On numerous occasions, I was struck by practical engagements with the modified arrangements. One night, I helped Felipe Borman heat coals on the barbecue grill in order to cook tapir steaks. The charcoal would not start. Felipe ran from the yard to the kitchen. He came back with a thick chunk of congealed tree sap brought from Zábalo, which he jokingly referred to as "Cofán kerosene." Just as he would with a Zábalo hearth, Felipe placed the sap among the charcoal chips and lit it to bring the fire to a self-sustaining intensity. The charcoal caught, but it did not burn sufficiently. Felipe ran back inside and returned with the lid of a large pot. He used the object to fan the fire with a series of quick wrist movements, just as his mother does in Zábalo. He told me that he looked for the feather fire fanner but could not find it. In the meantime, I had picked up a piece of sturdy cardboard that was lying next to the grill. I used it to fan the fire when Felipe got tired. I was surprised that he did not feed the flames with the scraps of paper that littered the yard.

After making sure that the charcoal was hot enough to cook the tapir, I stepped back and realized that Felipe was doing his best to reproduce the only fire-starting activity structure he knew. In Zábalo, people keep all the required items—sap, fire fanner, and pot lid—next to the hearth. Felipe failed to "see" that the objects that sat next to the grill in Quito—cardboard and paper—were also objectively available. Abstractly, he knew that he could use them. Nevertheless, in the breakdown of a routine learned in a Zábalo household, he turned to the only "recipe for behavior" (Schutz and Luckmann 1973:17) that he knew. His pragmatic response, however, made most sense in terms of a Zábalo, rather than a Quito, equipmental whole. Of course, there is neither much cardboard nor much paper in Zábalo, let alone barbecue grills. During my yearlong stay at the QCC, I witnessed many similar incidents.

Much of the color and content of the Cofán lifeworld derives from hunting and forest knowledge. At the QCC, Cofán children play the same games that they do in Zábalo: "hunting with a spear" (dodge ball), "woolly monkey" (climbing on trees and walls while others try to knock the climber down with balls or rocks), "collared peccary" (lying close to the ground until others see and give chase), and "capibara" and "tapir," which are simply hunting scenes acted out with some children as predator and others as prey. While sitting at the dinner table and eating stews of plantain and game meat, Borman often launched into elaborate hunting stories. He used stylized gestures and intonation patterns to describe how he killed the animal we were consuming. Quenamá and the children chimed in with their own experiences or stories from friends and family members. During the conversations, the children had a chance to demonstrate their forest knowledge. Borman used the opportunity to correct mistaken assertions or to add to the expertise of the youths, who were missing out on the practical enculturation of their lowland peers.

Speaking A'ingae is the most important practice that survives in Quito. The language is the primary means of communication at the QCC. Residents speak it while eating, playing, and helping each other with homework. The use of A'ingae relation terms is common. For example, Borman has a number of labels: *yaya* (father) for his children; *quindya* (older brother) for Quenamá's two younger brothers (who are also Quito students); *to'nto* (uncle) for unrelated Cofán students; and *Felipe yaya* (his teknonym) for the majority of visiting Cofán. When a large group of Cofán stay at the QCC for political meetings, everyone gathers in the living room to share news and stories. The house rings with laughter-filled A'ingae until the lights go out, and often throughout the night.

Linguistic, social, practical, and equipmental continuities alleviate the shock of residing in Quito. Most Cofán students, however, do not share their Quito quarters with their parents or other *naccu* members. Consequently, they are left without providers of discipline and comfort. Borman partially fulfills these functions, whether the children are related to him or not.[9] Nevertheless, he resents his expanding responsibilities as a "dorm parent." Over the years, his time and patience have diminished. The students realize that Borman is busy. Moreover, his status as a powerful *na'su* makes him a feared figure. In the words of one older student, "He is always busy, and we are always afraid of him." The fear, of course, is not unlike that of a shamanic apprentice. The children never worry that Borman will hurt them, but they do not want to pester or upset him. They realize that he is the *na'su* of the FSC. As such, he has significant power over their present circumstances and future possibilities.

When added to the interethnic difficulties of relating to Freddy Espinoza and Maria Luisa Lopez, the social tension that Borman inspires makes residence at the QCC even less *opatssi* for Cofán students. It would be an exaggeration to say that Cofán youths do not enjoy themselves in Quito. They eat well, they have fun with each other, and they know that a formal education expands their future options. Furthermore, they realize that their families and other Cofán people want them to finish their studies. Nevertheless, living in the city is a challenge. Cofán people's biggest fear is not that the children will lose their connection to a forest- and community-based way of life. Rather, they worry that the students will lack the courage and dedication that will enable them to stay away from their communities long enough to complete school and become *na'su*.

Cofán people place an immense amount of hope in the children who study in Quito. When I left the field at the end of 2002, the oldest had just entered high school. All the students voiced a desire to finish an undergraduate education in a national or overseas university. No one, however, knows what the youths will do or become. In large part, they base their assumptions on the life of Borman, who calls himself the "experimental prototype" for the Education Project. Nevertheless, never before have Cofán people tried to produce a Borman-esque figure through a collective project.

Participants, Gender, and Politics

The Quito students are not entirely representative of Ecuador's Cofán population. For one thing, nearly all are from Zábalo. While I was in the field, one child from Dovuno enrolled in the Tumbaco school. Within two years of my departure, Doreno and Tayo'su Canqque each sent students to the QCC. Although the FSC issued a call to all Ecuadorian Cofán communities, most people knew too little about the effort to participate. Nearly all of the original Quito students were related to Borman, Quenamá, and Aguinda. They included Borman's three children, one of Aguinda's children, two of Quenamá's younger brothers, and another Aguinda child who is distant kin to Roberto. Another student, Hugo Lucitante, is not close kin to any FSC employees. His father, however, was one of Zábalo's original residents, and he is a close companion of Aguinda and Borman. With its low cost, the Tumbaco school received a wider range of children. During my stay, many Zábalo families sent at least one child there.

In 2001 and 2002,[10] only one Cofán girl entered an outside "school." She lived in the town of Papallacta, where a missionary family offered to educate her in their home. She was a bright and relatively strong student. During her two years in Papallacta, she made substantial progress with English. Unfortunately, she returned to Zábalo after she "grew," or had her first period, and decided to get married. In addition, her father was very poor. He found it hard to part with the $25 per month that the Papallacta family requested. In general, however, the girl's gender made her failure less than a surprise. Although some parents are interested in sending their daughters to outside schools, Zábalo residents are much more likely to send boys away for an education. Their perspective reflects long-standing Cofán gender relations as well as a more recent history with Cocama in general and schooling in particular.

On the whole, Cofán gender relations are relatively egalitarian. Men and women are typically free to develop idiosyncratic preferences in terms of daily subsistence and social life. Nevertheless, women are at a disadvantage in certain domains, such as the acquisition of shamanic power, a process that reflects the cultural assumptions concerning schooling. I know of only one story that describes the prowess of a female *curaga*. She gained fame by battling a giant, insatiable anaconda, which no male shamans could lock up in its underwater abode. In Ecuador, I know only one living female shaman. Overall, the number of women with a long history of *yaje* consumption is small.

From the Cofán perspective, women have tremendous shamanic potential, but the reasoning behind this belief is built on a paradoxical fact. Because they menstruate, the *yaje* that they ingest fills their head, torso, and arms. It never descends past their waist, as their lower body is deeply contaminated by menstrual blood. With a restricted space, women become inebriated much faster than men, and they can "see" much more quickly. Nevertheless, the *yaje a'i* and other supernatural beings dislike the sight and smell of menstrual and pregnancy-related blood. Only exceptional women can overcome the challenges to become *curaga*. Often, the women most familiar with *yaje* are either postmenopausal or incapable of menstruating due to shamanic intervention.

Given the small number of women who have attained shamanic power, the Cofán associate a *na'su*'s qualities—boldness, aggression, and ethnic and moral ambiguity—with men. In the household context, people recognize that women can be just as *putsa'su* (boastful, angry, and ill-tempered) as men. Furthermore, most people agree that women participate in household decision making on an equal footing with their husbands. Nevertheless, all of my male and female consultants asserted that women are too timid to voice their opinions during interactions with important ethnic others, such as government and NGO officials. When I asked one woman why she could not become a *na'su*, she laughed and repeated a common refrain, saying, "I'm too afraid!" Women face obstacles in acquiring other-speech and other-knowledge, which makes it difficult to participate in domains of powerful difference. Although there are important exceptions, including Amelia Quenamá and Linda Ortiz, women are likely to remain rooted in household and community contexts.

Cofán people do not essentialize the avoidance of leadership as a quality of women in general. My male and female consultants shared tales of powerful women from other ethnic groups. They have witnessed vocal and knowledgeable Secoya women berating oil company officials in regional meetings. They have also watched violent Napo Runa women lash out at men in intercommunity meetings concerning territorial boundaries. Zábalo residents realize that non-Cofán women are just as capable of outside education and leadership as men.

Some Cofán people suggest that Cofán women are less likely to become leaders because of their specific history with Cocama men. During the early years of oil exploration, Cofán women were afraid of being chased and raped by Cocama workers. When Cofán territory was colonized in the 1960s and 1970s, Cocama settlers propositioned and threatened Cofán women. The absence of peaceful interactions with Cocama put Cofán women at a disadvantage with regard to learning Cocama language and social norms. During an interview, Borman related an incident in which Linda Ortiz explained the position of Cofán women to a Cocama NGO worker who was conducting an inquiry into the Cofán social and political situation:

> The woman who was doing the investigation questioned us very pointedly. She was going through all the spare-time activities—who made the decisions and all of this sort of stuff—and she found out that she was looking at a very egalitarian bunch. So she asked Linda and Roberto, "Well, who's the boss in your family?" And Roberto laughed and said, "She is." And so she turned to Linda, and Linda just glared at Roberto and said, "He doesn't always obey me though!" And then the woman asked, "Well, then why aren't there more Cofán women in leadership positions?" And Linda said, "Because your men won't let us be. Your men won't let us even learn Spanish. Your men proposition us. They want to sleep with us. If we try to learn Spanish from the women, they're so jealous that they think we're going to try to get their men. There's no option for us to learn Spanish." With regard to education, once again the attitude has been that girls aren't going to be the ones who are handling the outside world.

A number of other factors explain the unequal participation of women in outside schooling. As was the case with the girl who went to school in Papallacta, women's earlier physical and sexual maturation means that they are less likely to finish their studies before deciding to take a spouse. Many female *dusunga* begin to look for husbands from the age of twelve or thirteen. Male *dusunga* normally wait until they are seventeen or eighteen, the age when a student ideally completes high school. Moreover, all Cofán accept that distant students might begin romantic relationships with non-Cofán people. The possibility is gendered in important ways. Historically, Cofán women were more likely to enter sexual relations with ethnic others, especially

Cocama. Cocama men desire Cofán women as sexual objects and potential spouses. Cocama women, in contrast, show little interest in Cofán men.

In the most intensively colonized communities, such as Sinangoe and Doreno, a number of Cofán women have married Cocama men. None of the inverse cases exist. During the last thirty years, Cofán people have become cautious about Napo or Cocama marrying into their communities. They worry that the ethnic others will disrupt preferred modes of sociality. Because boys are less likely to marry into a non-Cofán community or to bring a non-Cofán spouse back to their village, Cofán people consider them to be less of a risk for outside schooling.

Finally, a number of contingent factors explain the greater number of male students. The most likely entrance point for outside education continues to be the Tumbaco school, which is only open to boys. In addition, demographic contingencies create a situation in which more Zábalo boys are candidates for outside schooling. In the community, one couple has a large number of daughters and some income. The father, however, is a known "prohibitor." He jealously controls the future of his lone son and his daughters, whom he wants to keep in his *naccu*. Another couple has a large number of daughters but no income. They cannot supply the $30 per month that the FSC requests for all students. All of the participating families, in contrast, have large numbers of boys. Borman and Quenamá have no daughters. Quenamá's younger siblings are all boys. Aguinda and Ortiz have only boys. And Hugo Lucitante has four brothers and one sister, who is too young to enter school.[11] In the future, I believe that more girls will participate in the Education Project. Nevertheless, they will have to overcome the challenges posed by demographic chance, cultural expectation, and financial circumstance.

Fears and Hopes

In general, Cofán people are not afraid that Quito students will lose their identity and stay in the city. If they do acknowledge the possibility, they assert that the Education Project is a political necessity. Moreover, they claim that it is impossible to predict anyone's future. As one parent stated, "How would one know? We can't know the future." The only evidence they have is Borman. Cofán people realize that much of his political capacity stems from his multicultural development and formal education, which did not destroy his commitment to Cofán life.

Cofán voice various opinions about cultural change. Most of my Zábalo consultants could not think of a single person who left the *tsampi* to live in the city. A handful of Sinangoe women married Cocama colonists, and in-marrying Napo virtually took over Dovuno. In addition, a small number of men and women have married into Siona, Secoya, and Napo Runa communities. Nevertheless, the Cofán view these cases as exceptions, and they do not understand them as the intentional abandonment of one's Cofán-ness.

Some Cofán communities have begun to experience forms of social change associated with young people's increasing interaction with Cocama. Borman and Aguinda told me that in Sinangoe, where intermarriage with Cocama is the most common, Cofán residents did not speak A'ingae in community meetings until recently. When Borman, Aguinda, and Quenamá first began to work in the community, they only spoke A'ingae. Consequently, the quiescent Cofán reentered the conversation, and the Cocama were left out. In addition, Quenamá told me that simply witnessing the Borman children's multilingual skills taught Sinangoe residents that their young people could learn other languages without forfeiting A'ingae or Cofán identity. In Doreno, the problems are not as severe. Some adults, however, suggest that children who attend school in Cocama towns are becoming "fools." Reportedly, they say offensive things in Spanish, and some refuse to speak A'ingae in Lago Agrio.

Aside from changes in language and character, people worry that sending children to outside schools will complicate their socialization. The concern, however, is not serious. People expect the children to make up for lost time when they return to their communities. Summer and winter vacations represent opportunities for ensuring that the students do not lose their ability and desire to live a forest-based life. Zábalo residents claim that the children are glad to visit. They enjoy hunting, fishing, and eating the foods that they miss while in Quito. During their stay, the students face the same kind of social pressure that Borman confronted while on break from Limoncocha and Quito. If they incorrectly identify a species or disembowel a game animal from the front rather than the back, for example, all present joke with them for their mistake. The responses are a good-natured means of letting the children know that forest life, too, depends on important forms of knowledge and practice.

In addition, Quito students look forward to visits to home communities as opportunities for romance. In Doreno and Zábalo, there are plentiful opportunities to talk to, laugh with, and make lovers of Cofán girls. The possibilities, however, can be problematic for educational success. Although they anchor youths in community life, they threaten students' commitment to finishing school. The presence of a willing marriage partner in Zábalo or Doreno can make the isolation of Quito too much to bear.

The potential transformation of Quito students is an ambivalent possibility. The defining characteristic of *na'su*, after all, is their ability to utilize their ethnic and moral ambiguity to deal effectively with social and supernatural otherness. Cofán students inevitably incorporate some of the practices and desires of external domains. The process, however, need not result in a true displacement or a permanent change. While talking with Aguinda about the risks, he shared his reflections on his own transformation while residing away from Zábalo and Doreno:

> Let's say that I go to the United States. I go and I stay there for ten years. My customs won't truly change. Look at yourself. You came from the United States to live here. You went to Zábalo. You did that and you don't really think that you're Cofán. But you went and you resided there. You went and stayed there for a year and you ate things without concern. You ate anything, even the food of cockroaches [laughs]. But you didn't think, "Yes, I've recently become Cofán by eating things like this." And when you go back to your home, you'll become Gringo again. That's the way that *a'i* customs are, too. I can stay in Quito. Whatever people feed me, I eat. But I'll go back to my home, where I speak A'ingae with my own people. I drink banana drink and manioc beer. That's all I do.

There are a few cases in which an individual left a Cofán community yet continued to maintain a connection to Cofán people. For example, a former president of the Cofán ethnic federation, who now lives in Lago Agrio, was never a good or happy hunter. Instead, he focused on teaching and political activities. He continues to work for FEINCE, and he is active in the suit against Chevron. Even though he is not capable of practicing a forest-based way of life, he is proud of his language and identity. Moreover, he shares his home with Cofán people who travel to Lago Agrio for political, medical, and economic reasons.

Another man, born in Doreno, is only marginally Cofán. He speaks no A'ingae and has not lived in a Cofán community for years. As of 2012, he resides in the distant town of Puyo. For years, he occupied a leadership position with a regional indigenous federation. Although Cofán people question his identity, they appreciate his effort to bring aid to their communities. He gives them reason to believe that if students do not return to their communities, they still might become important supporters of Cofán projects.

Other factors mediate anxieties concerning the students' return. Zábalo residents stress that sending children to outside schools is not supposed to produce a total transformation of the Cofán nation. Instead, they view it as an effort to generate a small group of knowledgeable individuals. If enough children return to work as *na'su*, people will be much less concerned about the possibility of a few students losing their commitment to friends, family, and the wider Cofán community. When I asked Aguinda what Cofán would think if a student decided to work for himself rather than his people, he replied:

> That would be really bad. They wouldn't be worth anything. But if the community already had a lot of people who had learned, it would be fine. But if there were none, like now, and if one or two children learn in school and then leave forever to work, that would be bad. They wouldn't be thinking about their territory or their people. But if around twenty already came back to reside in Cofán communities, it wouldn't be so bad if others decided to go somewhere else.

Most people do hope that the students will help the Cofán nation. In conversations with parents, I did not encounter expectations of personal, household, or family gain. Instead, people spoke of the children as objects of community and national interest. No parents stated that their children owed them a "repayment." One Zábalo man described his counseling of his child in the following terms:

> I really don't want my child to learn just for himself. I want him to learn so that he helps his community. His father [the speaker] struggled so that he could learn, so that his child could help his land so that in the future all will become well. That's what I want. I say to my child, "After you learn, you have to help your land. Look at me. I don't speak Spanish well. I can't help my own community. Now, Randy is helping. He has compassion for *a'i* and he helps. He speaks to the Cocama and he helps the land, so that *a'i* have land to live on." I explain all of this so that he thinks and understands. If he doesn't, he won't be worth anything. If he goes and learns and stays to work in the land of the Cocama, how will he help his own land? That would be bad.

Cofán people hope that the students will be able to do many things. They want at least some of them to be *na'su* who can speak Spanish, English, and other European and indigenous languages. They want them to interact with government officials so that they can secure Cofán territory and obtain services for Cofán communities. They want them to engage national and international NGOs in order to bring more *proyectos* to their villages. They want them to reestablish Cofán tourism. They want them to be true teachers for the children who cannot leave their home communities to get an education. And they hope that the students will take on new roles, such as doctors, lawyers, engineers, and mechanics. As Aguinda stated:

> I really get sad about that. *A'i* still lack a lot. Everything, like engineers, doctors, and lawyers. I'd like the *a'i* who have completed their education to know that and to become that. So that if there's some work that needs to be done in *a'i* territory, *a'i* themselves can do it, so that it's theirs. And if *a'i* become doctors, then *a'i* themselves can begin to cure sick people. And if *a'i* become lawyers, then *a'i* themselves can work to fix our problems.

Below, I briefly consider the cases of three Cofán youths who have progressed and faltered on their educational paths. None of them have reached their ideal position, but all have experienced unique opportunities and challenges. Although some of the details appear to be "sensitive," I feel that including the names of the students is essential to describing their situation. All are closely related to key players in the Cofán experiment. At the time of this writing, they are all legal adults, and they have reviewed my account and given their approval.

Felipe

Born in 1988, Felipe is the oldest son of Borman and Quenamá. He attended the Zábalo school through third grade. In 1998, he entered the Alliance Academy. Although he grew up speaking English and A'ingae, he had language difficulties during his first years in the city. His grades improved after he became proficient in English, but Felipe never became a straight-A student. I met him in January of 2001 when I arrived at the QCC. From the beginning, I had a tense relationship with Felipe. He constantly teased me. I had just settled in with the Borman family, and I did not want to jeopardize my position by attempting to discipline him in any way. My immediate assumption was that he suffered from a case of being the "chief's son." I hate to admit it, but I felt disheartened when I thought of him as a potential leader of the Cofán nation.

After I spent time in Zábalo and learned A'ingae, Felipe let up on me. I became convinced that his early demeanor was simply the reaction of a feisty thirteen-year-old who wanted to defend his turf from a questionable newcomer. As he got older, I realized how proud he was of his Cofán identity. I discovered his love for hunting and forest life. On the first day of my initial five-month stay in Zábalo, he took me through the *tsampi* to teach me how to shoot tinamou. The next day he went on an extended hunting trip. He returned with a couple of good-sized collared peccaries. When I interviewed his Alliance Academy teachers, they told me that he wrote about hunting and fishing in many of his assignments.

Of all the Cofán schoolchildren, Felipe was the most vocal about his identity. Nevertheless, his teachers told me that he was just "one of the guys." In 2001 and 2002, he blended in well with the other students, who also wore baggy pants and baseball hats. During return trips in 2003 and 2004, I discovered that Felipe had made good use of his nappy hair to cultivate a countercultural-looking mane. (His choice made me think about Borman's early brushes with the school administration for growing a moustache.)

Many of the Alliance Academy students are impossible to differentiate from average North American teenagers. There is a broad acceptance of the physical self-stylings of such youths as Felipe. Like his classmates, however, Felipe was (and is) a Christian. His Bible class teacher told me that he was one of the best students, always bringing a creative edge to his arguments. I saw other aspects of his creativity while living at the QCC. In a situation similar to the fire-fanning incident described above, I once looked out of my window to see Felipe manning the barbecue grill. Once again, he could not get the coals to burn. Rather than tire himself out with the pot lid or feather fanner, however, he found an extension cord and plugged in a vacuum cleaner. He switched it to reverse and blew air onto the stubborn charcoal. The incident made me think about the knack for experimental thinking that has been so important to Borman's work.

Felipe always told me that he wanted to marry a Cofán girl. He found it hard to imagine a Gringa or Cocama who would be able to handle life in Zábalo. He sometimes spoke about pursuing a well-paying job in Ecuador or the United States. At other times, he talked about finding a way to earn money in Zábalo. In 2001, I asked Felipe whether going to school in Quito made him different from other Cofán. In his reply, he talked about various possibilities. The relevant factors were money, political aid to the Cofán nation, and caring for his father:

> Well, I am kind of different because I'm going to school and I'm trying to get educated so I can help the Cofáns when I grow up. That's what I want to do. But first I want to be a lawyer and make some money [laughs]. But while I'm a lawyer I want to help the Cofáns somehow. I might go to the States and be a lawyer there. But when my dad gets old and he needs a lot more help, I'll come down and begin to help him here. And then I might take over my dad's stuff and get something good, so that I can make money doing Zábalo stuff. I might do a big thing for tourist attraction. Then I'll make money, and Dad

can sit back because he paid for all the stuff when I was growing up. He paid for my education, so it'd be time for him to rest and let me do all the work.

In 2006, Felipe entered Knox College in Galesburg, Illinois. Knox is an excellent liberal arts school with a high proportion of international students. In the following years, I saw him a number of times in the United States and Zábalo. For Felipe, college appeared to be a lot of fun. In 2008, I spent a month working with him and two other Cofán students on a project at the Field Museum. In 2009, he drove with his Nepali girlfriend from Galesburg to the University of Iowa to listen to me give a talk about his father and Cofán politics. In June of 2010, he graduated from Knox with a degree in political science and returned to Ecuador. When I visited Zábalo in July of 2010, a rumor circulated that Felipe had married a Cofán woman from Doreno, but no one was able to provide confirmation. I did discover, however, that he had taken charge of the construction of the Cofán park guard headquarters in Lago Agrio. He is now one of the main coordinators of the program, which demands that he move constantly through Quito, Lago Agrio, and the most remote portions of Cofán territory.

I am Facebook friends with Felipe. He lists no current residence or occupation. His profile picture is a recent image of him, shotgun in hand, crossing a log bridge over a stream in Zábalo's *tsampi*. The title of his first photo album is "Work, Conservation, Life." It contains pictures of forest camps, Cofán park guards, and his family. The first line of his "Bio" contains three words: "Jungle of Ecuador."

Nivaldo

Born in 1982, Nivaldo Yiyoguaje is one of Amelia Quenamá's younger brothers.[12] He spent his early years in Campo Eno, the Siona community of his father and paternal uncles. Nevertheless, he stayed in his mother's home community of Doreno for long periods. When his mother and father joined Borman and Quenamá in the early 1990s, Zábalo became his home. Nivaldo speaks A'ingae and Siona, and he is also one of Zábalo's best Spanish speakers.

Nivaldo completed most of his primary education in Zábalo. He was part of the first group to attend the Tumbaco school, from which he graduated in sixth grade. He then moved to the QCC to attend the Colegio del Futuro for two years, after which he enrolled for two years at the Colegio Francés. When I first met him, in January of 2001, he was eighteen and in seventh grade. The poor quality of his Zábalo education meant that his school position and biological age did not match. Like most Cofán people, he is fairly short. His size helped him to fit in with his classmates, although most students realized that he was significantly older.

Nivaldo had a harder time in school than the other Cofán students. In the summer of 2001, he could not return to Zábalo for vacation because he had to make up failed classes. I found him to be a smart, sweet, and always laughing *dusunga*. Nevertheless, he often lacked the discipline to complete homework or do well on tests. Most of his Colegio Francés teachers professed a strong fondness for him. They described him as reserved, quiet, and *tranquilo*, even though he was an accomplished athlete.

Unlike the other Cofán students at the Colegio Francés, Nivaldo was willing to talk about being Cofán. He was the only one who wore his tunic and headdress for a school program on Ecuador's *culturas*. Although Nivaldo was willing to become a cultural object for his teachers and classmates, I always thought of him as the most *opa* QCC student. Many of his Siona relatives are accomplished shamans. Nivaldo told me that he is like his father, who could not progress on the shamanic path because of the suffering and prohibition on romantic activities. At the QCC, he compared shamanism to *estudio*. The association made me question his potential for completing his education.

When I arrived at the QCC, Nivaldo said that he was in Quito "in order to finish studying, so that I can help my community, so that I can help my family and the whole community." Nevertheless, he was uncertain about the expectations. In large part, he modeled himself after Borman. He said that he wanted to have a house in both Quito and Zábalo, like Borman. In addition, he told me that he wanted to work for the FSC, to become a teacher, or to learn English and become a tourist guide, "whatever Randy is."

Many times, Nivaldo told me that he did not want to be a *na'su*. Nevertheless, he said that he would enter a leadership position if the community told him to. More than the other QCC students, Nivaldo was aware that becoming a *na'su* requires directing others and becoming an object of bad talk. For him, neither was an attractive possibility. When I asked if there was something else he could do, he shrugged and replied that he could simply be an *a'i* and reside in Zábalo. He took a big step in that direction by marrying a Zábalo girl during his 2002 summer vacation. Just before the event, he told me that he was upset that people wanted him to delay the marriage so that he would not jeopardize his studies. At the age of twenty, he had just completed eighth grade. The prospect of four or more lonely years in Quito appeared to be an unbearable burden.

After his marriage, Nivaldo returned to Quito for another year of school. Within a few months, however, he retreated to the lowlands. He tried to convince Quenamá and Borman to let his wife live with him at the QCC, but they decided that it would set a bad precedent. When I spoke to Nivaldo's brothers in Zábalo, they expressed sadness and anger at his choice. They felt that he had little to show (i.e., no degree) for four and a half years of outside schooling. In addition, significant tensions emerged between Nivaldo's birth family and his in-laws. The former blamed the latter for allowing the marriage and ending Nivaldo's education. Determined not to let Nivaldo's education be a waste, Zábalo residents pressured him into becoming the community schoolteacher. After a short time, however, he decided that the job was not for him.

Nivaldo reserved most of the blame for himself. He simply lacked the discipline to *vanamba atesuye*. His discomfort with the idea of becoming a *na'su* and his uncertainty about his future exacerbated his poor motivation. Borman was his only role model. Nevertheless, he did not see himself as desirous or capable of taking on a Borman-like position in Cofán society. Borman uses Nivaldo's story to convince potential donors that their main fear concerning the Education Project—that Cofán students will abandon their communities to pursue an urban, commodity-rich life—is unfounded. The real challenge, Borman explains, is keeping children away from Cofán communities long enough to complete their education.

Even though he did not receive a degree, Nivaldo's experience with outside schooling represents an important political resource. Aware of his good Spanish and substantial experience in Cocama settings, community members elected him president of Zábalo for one year. More importantly, he decided to enter the park guard force. Over the next few years, he became one of the program's coordinators. In 2005, the Field Museum sponsored his travel to a national park that they helped establish in Peru. They wanted him to share Cofán experiences of environmental management with the new reserve's inhabitants. Subsequently, the FSC sent him to the United States to participate in similar exchanges. Although Nivaldo does not appear to be cut out for long-term residence in Quito, I believe that he will continue to be an important actor in Cofán politics.

Hugo

Born in 1987, Hugo Lucitante is the oldest son of Bolivar Lucitante, one of Zábalo's first residents and community leaders.[13] Hugo's biography is the most unusual story of Cofán involvement with outside education. As an undergraduate, a Seattle woman named Miranda Detore made a ten-day visit to Zábalo as part of an "adventure school." While traveling in lowland Ecuador, she saw the environmental devastation caused by the oil industry.

Meeting the people of Zábalo, in her words, "profoundly touched my soul." She returned to study abroad in Ecuador in 1997. On the advice of Borman, who wanted her to get a "feel" for Cofán culture, she lived with the Lucitante family for two months.

Detore's experience in Zábalo strengthened her commitment to aid the Cofán nation. She offered to become the legal guardian of Hugo so that he could return with her to Seattle, learn English, and get a high school education. At the age of ten, Hugo moved to the United States. From the beginning, he knew that Detore and his family wanted him to become a leader of the Cofán people. Explaining his parents' choice to let him go overseas with a near-complete stranger, Hugo said, "They wanted me to have a good education and to come back here and help the Cofán."

Hugo has attended seven different schools: four years in the Zábalo school, fourth grade at a private Seattle school, fifth grade in a public Seattle school, sixth and seventh grades in a different private Seattle school, eighth grade in the Alliance Academy, four years at a Catholic high school in Seattle, and a semester at the Universidad San Francisco de Quito. In 2002, he told me that his grades were "Okay, but not great." Hugo said that at each new school he presents a slide and video presentation on Zábalo and Cofán culture. He sometimes speaks about the social and environmental consequences of petroleum production. To help fund Hugo's education, Detore brought tour groups to Zábalo. Given his fluency in A'ingae and English, Hugo acted as one of the main guides. While in Seattle, he lived a fairly normal teenage life. He played soccer, went to movies, and rode bikes with his friends. He is a good-looking, sweet, and soft-spoken young man. He had no problems making friends in either Seattle or Quito. When in Zábalo, he loves to hunt, fish, and play volleyball.

Hugo has consistently voiced his desire to return to Zábalo after finishing his education. Many Zábalo residents point to him as proof positive that outside schooling does not threaten the commitment of Cofán children to a forest- and community-based way of life. While I was conducting an interview with Hugo in Zábalo (in English), a young A'ingae speaker listened. When I asked Hugo if people thought he had changed, he asked the onlooker, who laughed and said, "He makes language mistakes sometimes. Sometimes his language is like that of a Gringo."

It appears that no one in Zábalo feels that Hugo's future is in doubt. Hugo's father is fairly strict. He told me that he constantly counsels his son about his future obligation to help the Cofán. During an interview, Hugo explained that people in the United States always ask him what he will do with his life. He always says the same thing: "I want to live in Zábalo. It's my home. I feel comfortable here. And my family, everything. Since I was little, when I had the chance to go to the States, I wanted to learn, come back, and live here."

Although Hugo sometimes seems too *opa* to become a truly effective *na'su*, he has proved his dedication. Early in life, he began to take on political roles. With the help of Detore, Hugo spoke in front of the United Nations in New York City about oil contamination in Cofán territory. Detore flew his parents up for the event. According to Hugo, their response to the United States was the same as his: "Nice to visit, but you can't live here." In 2003, the Field Museum paid for Hugo to travel to Chicago, where I worked with him on public-speaking events about Zábalo's conservation projects. In 2005, the Field Museum brought him back to Chicago, where he did a summer internship with the departments of botany and zoology and the ECP. In 2007, 2008, and 2010, Hugo worked with me on the "Cofán Historical Mapping Project," a Field Museum–supported initiative that we first envisioned during his 2005 visit to Chicago. In 2007, a documentary crew from Seattle began working on a film that uses Hugo's story to personalize the struggle against oil-based development in Amazonian Ecuador. Entitled *Oil and Water*, it is still in production.

In early 2009, Hugo married Sadie Kirk, a young North American woman he met in Seattle. After working in the United States for six months, they decided to move to Zábalo. Even though Sadie has yet to master A'ingae, they immediately became community schoolteachers. I was worried that Sadie would not be able to handle life in the forest. When I visited in July of 2010, however, they had constructed their own home along the river. It appeared that a year in Zábalo had treated them well. Nevertheless, both wanted to return to the United States to finish college. After much discussion, we decided that the city of San Antonio, which is also my home, would be a good starting point. They lived with me for a month before finding work and moving into their own apartment. Half a year later, however, they became disillusioned with menial wage labor and felt the pull of "home." They returned to Zábalo and built a new, larger house. After completing another round of filming for *Oil and Water*, they decided to move back to Washington State, where they have residency and can enter state universities without a year-long waiting period.

Hugo does not yet know what he wants to do as a career. His time at the Field Museum increased his interest in environmental science. His knowledge of English and experience as a guide make tourist work a possibility, too. Hugo claims that he wants to do anything that will enable him to spend significant amounts of time in Zábalo. As with Nivaldo, Hugo is

Hugo Lucitante working on the Cofán Historical Mapping Project (Photo by Michael Cepek)

not sure that he is the right kind of person to become a Borman-like *na'su*. He does not consider himself to be a natural leader. From my perspective, his shyness would sit uneasily with a job as the Cofán nation's central spokesperson. Nevertheless, he wants to help his family, his community, and his people. Recognizing his skills and determination, the Cofán elected him FEINCE's Director of Community Development shortly before he moved to the United States.

NECESSITY AND FREEDOM

Borman understands the risks and complexities of the Education Project. He realizes that studying in outside sites is a true challenge for Cofán children. In addition, he is aware that some students might change drastically while interacting with Gringo and Cocama for years on end. Nevertheless, Borman shares the opinion of all Cofán people, who know that there is no other option if the Cofán are to survive the dangers and reap the benefits of the global era. One day, other leaders must replace the gringo chief.

Borman views the Education Project from the particular perspective of his own biography and political vision. He is not willing to think of future Cofán leaders as anything less than the best candidates for their "job," which is to protect, manage, investigate, and profit from the most biologically diverse territory in the world. If the Education Project is to succeed, Borman argues that the students should view their work on behalf of the Cofán nation as a free choice rather than an obligation. He wants them to know that their education will allow them to do or to be anything they desire. Only true freedom, he argues, will ensure their commitment.

Borman is critical of many indigenous leaders, whom he describes as "neither here nor there." He feels that the majority of ethnic federation officials are products of a poor education system. According to him, mediocre schooling has cut their ties to their home communities while failing to prepare them for anything but careers as professional spokespeople. They have no choice, he asserts, but to request money from a world that views native Amazonians as objects of sympathy. Moreover, from his perspective, they do not have the skills, knowledge, and experience to imagine and implement new forms of political transformation. When I asked Borman what will happen if none of the QCC students succeed, he shared the following prediction:

> First of all, I see us being stuck at what I have called the professional Indian leadership category. Our best example of the professional Indian leadership category would be Alonzo.[14] He is the archetype. He doesn't speak the language. You could drop him in the middle of the jungle and he couldn't find his way out with a compass and a GPS included. He might be able to find his way out if he hears trucks in the distance. His understanding of which end of the blowgun to blow through is strictly limited. Calluses on his hands are largely from working in his coffee fields, if he has to get any money at the moment. But still, he considers himself a Cofán for political reasons. Here's a person who does not understand his community, has never lived the life of the majority of his community, does not speak the language of his community, is married to a member of another community, and has very, very few ties. He has a very poor educational profile, but he's on the board of directors of the Coordinating Body of the Indigenous Organizations of the Amazon Basin. And he's out there wandering around selling his programs in the name of his indigenous constituents. I think we'd get a lot more of that class out of the education system as we see it now. [Is there something about the education system that creates a person like that?—M. Cepek] It's the whole process of mediocre training at all levels. There's really nothing else he can do. He doesn't really have any other job open to him. He could become a teacher, but he doesn't want to. He's not a teacher.

Borman's statement points to the specific challenge of the Cofán experiment as an indigenous political project. All Cofán *na'su* are ethnically and morally ambivalent from the perspective of the people they lead. The problem with the individual Borman describes, however, is that his level of knowledge and power relegates him to a position in which he is minimally helpful to the Cofán nation for two reasons. First, he does not understand the life situation of the majority of Cofán people. And second, he is incapable of knowing or fighting the Cofán nation's real enemies. For Borman, the most significant challenge facing Cofán people is not only wide-scale forest destruction. It also includes the dominant structure of NGO politics and the problematic operation of transnational networks of science, conservation, and sustainable development. Borman believes that Cofán activists must creatively transform these forces if they are to overcome the problems that face their nation.

Based on his own negotiation of political crises, Borman believes that the only possibility for Cofán survival is to create leaders who share his value orientation and skill set. According to his perspective, only individuals with the best formal education will succeed in managing the complexities of indigenous and environmental politics. Only individuals who share the orientation of a *curaga*, rather than that of an *opa a'i*, will be bold, aggressive, and determined enough to sacrifice the joys of community life to engage in long-term political action. And only those individuals who are familiar with a forest-based way of life—as well as capable of freely choosing to participate in that way of life—can be trusted to use their ambivalent powers in ways that benefit the Cofán nation. When I asked Borman what kind of "pull" future leaders would feel toward Cofán communities, he replied:

> I think there would be a very strong idealistic pull back. I felt very strongly about it. I wanted to learn how to swing a machete. And I wanted to excel in Cofán culture because the educational culture that I had been in gave me that option. It was an option, not a necessity. If I couldn't do anything else because I didn't have the training, I would've fought against it. But the fact that I could—I had the doors pretty much open to me—made it an option for me to return and learn how to swing a machete adequately. I hadn't learned that when I was growing up. I had learned and very much loved to hunt. But I would say that that profile fits most of these kids. All of them. The hunting, the freedom of the forest—all of that stuff is very, very attractive. But getting down to the nitty gritty of life, where you have to go out and build a house and all of this sort of stuff. Unless you have a very strong idealistic pull on top of the straightforward necessity, you're going to try to get around it. It's hard work. It's hard, sweaty work. And unless you've gone far enough in your idealization of that lifestyle, it's very hard to stick with it. But I think the very activity of being outside of that—knowing that you're Cofán at heart and by heritage—is an active way of idealizing that and creating an internal mythology of what that is about.

Here it not only means this idealized picture, but it's also where you've got your family. It's also where you've got your ancestral ties. This is where the real pull is occurring—it's not only an idealization. And I don't think you can get to that point until you have options. When you're just dealing with how you are going to survive, there's no room for those options. Idealism is a luxury. That's basically what I'm saying. And it's a luxury that's only going to be available to people if they're fully educated and therefore fully capable of handling whatever they want. My way of life, my return to the Cofán world, was taking advantage of options. I created a situation where I really can't escape from it very easily. I can, but it's not the easiest way out for me. But it was definitely my creation. It wasn't a situation where I had no choice in the matter.

The deepest Cofán explanation for Borman's political success is his particular value orientation, which combines aggressive ferocity with collective commitment. If Cofán students are to reproduce Borman's abilities, they will have to share his desire for adversity, struggle, and overcoming. Moreover, they will have to commit themselves to the welfare of a Cofán nation that does not embrace their basic impulses. In other words, they will have to transcend the opposition between the jaguar and the woolly monkey, or the *curaga* and the *opa a'i,* if they hope to successfully negotiate a troubling future.

Most Cofán people still do not understand what capacities are necessary for such a task. Nevertheless, they realize that leaders such as Borman are essential. They recognize that Borman's desire to live as a Cofán and for the Cofán is a choice. Ultimately, their awareness of Borman's commitment provides the strongest foundation for their faith in the Education Project. If Borman can leave them but chooses not to, they wager, why would their children choose anything different?

Conclusion

"MARIANE CÓNDASE'CHO" ("THE STORY OF MARIA")

In the midafternoon, on a path close to a garden, a girl walked by a tree. "Where are you going?" someone said to her. "Who just spoke?" replied the girl. She stopped and looked behind her. Again, the voice spoke, "Where are you coming from?" She was scared and went to look at the tree trunk. Someone was sitting inside. The girl asked, "Why are you calling me?" The person in the tree said, "I just want to know where you're coming from." The girl asked, "Why do you live in this tree?" The person simply replied, "This is where I live."

The girl returned to the village and told some old people what happened. She asked them, "What person could be in that tree trunk, with white clothes and white skin, sitting in a nice bed? Who was that person who just spoke to me?" The old people thought and said, "What could be there? Let's go see."

One old man went to look. He entered the tree trunk and sat down. Maria [now named] asked him, "Are you all living well?" The man replied, "Yes, we're living well. Where did you come from?" Maria said, "Our father left me here. That's why I'm here. He told me, 'Now you have to take care of this place and these people.' That's what he said, our father who lives in the sky. I'm here to take care of you all, so that you don't die, so that the *cocoya* don't come. That's why I'm here." The old man listened, and he understood.

The man went back to the village and told all of the younger people that Maria lives in the tree trunk. Everyone listened, and they went to see Maria. The young people went and looked, and Maria saw them and smiled and laughed. She didn't speak to the young people, but to an old person she said, "Build me a house. I want to live in a house here."

They built her a house, and Maria lived in it. Our father from the sky left a book in that house and then left again. Maria made a school, and the young people studied in the school and read the book. They then said, "Our father is coming for Maria. He wants to take Maria away. But Maria doesn't want to leave, because if she leaves, we won't be able to live." So the people decided to build another house, a nice house for Maria to live in.

While Maria lived there, not one child died, and there were no sicknesses. The people reproduced well, and the population grew.

Then, some priests from Colombia came and saw all of the Cofán. "How many are there?" they thought. They wanted to take the Cofán away to make wars in Colombia. The people didn't want to leave, so they stayed. The priests returned and forced the Cofán to build a house. They told them it was for the soldiers. The Cofán built a house to make them happy. They built a big house with palm-wood walls. The priests stayed there.

On Sunday, the priests called the people to the house to have a meeting. All of the people went, even the children. Not one person remained in the village. They all gathered inside, and the soldiers came to the door. No one could escape, and with machetes the soldiers cut and killed many people. That's how the priests did it.

Afterward, Maria was left with one very old couple, the only ones who survived. The Spanish didn't kill those two. They met together and said, "How are we going to survive? Let's go to the other side of the [Aguarico] river." They went to the other side. They built another house on the other side of the river and brought Maria there. They stayed there for four days. On the morning of the fifth day, they went to see Maria, but she wasn't there. "Where did she go?" they thought.

They decided to return to their old home of Bifeno [an Aguarico village site] to see if Maria was there. She was there in an old, rotting house. They wanted to bring her back to the other side of the river, but Maria said, "No, I want to stay here. This is the land that I want to live on." Then, the old man said, "We're going to live here, too. I'll clear the land, and I'll keep you here." So he kept Maria with them. They stayed there, and once again, the children reproduced, and the population increased.

The priests returned three times to kill the Cofán. The priests massacred the Cofán from the Aguarico River three times, and they massacred the Cofán from the San Miguel River only once. Because they were massacred so many times by the Spanish, the Cofán of the Aguarico speak slowly and in a soft voice. The people from Santa Rosa [on the San Miguel River] speak differently. You can't understand them. They speak in such a loud and strong voice. When they shout, you become frightened and almost die!

The Cofán reproduced again, and the Colombian priests came from Pasto. They heard about Maria, and they came to find her. They came, saw her, and stole her—the a'i atapaje'cho [Cofán-reproducing thing]. The priests brought Maria to Pasto and built a nice house for her.

Some Cocama told the Cofán that Maria was in Pasto. The Cofán understood, and they went to find Maria. On a house, there were words. Some of the Cofán could read, and they understood the words. "Ah, here's where Maria is," they thought. They went inside the house, but Maria wasn't there. It was a nice Cocama house, painted really well. The Cofán went in and looked, but they just saw an anaconda inside. Then the Cocama asked them, "Have you seen Maria?" The Cofán replied, "No, we didn't see her. We only saw the anaconda." The Cocama said, "There's no anaconda. Why did you see that? You're not supposed to see an anaconda. You're supposed to see Maria."

The Cofán understood and went to the priest. The priest prayed and baptized them. He said, "Now, go and look. You have to think well of God and look." The Cofán agreed, and they went to look. They saw Maria sitting there. Maria said, "Why did you come?" The Cofán said, "We were sad, and we wanted to see you." Maria asked, "Where are you from?" They answered, "We're from far away, from San Antonio [on the Guamués River]. We came to see you." Maria took the hand of a Cofán, and put her other hand on his head, and said, "I am living here."

Then, she said, "Look at this book that I have." The book was close to Maria. The Cofán looked at the book, and the book explained how many dead people there are, and how the world will end. It was written by God. One Cofán could read it. He read, and he understood. Then, Maria said, "Are you going to come and visit me again?" The man said, "Yes, I'm going to visit you. All of my family wants to see you."

On Sunday, he went with his family and arrived at the house. The man that saw Maria before saw her again, but his kinspeople could not see her. It was just a *cocoya* there. But if a person is a Christian, he can see Maria. Maria is still there.

SUCCESS AND ITS CONSEQUENCES

I recorded "The Story of Maria" with Atanasio Criollo, Zábalo's oldest resident, in 2002. It describes the origin, power, and destiny of a supernatural figure who protected the Cofán nation from death and destruction, allowing them to become a strong and numerous people. Despite its portrayal of a white-skinned messiah ambivalently associated with a quasi-Christian god, the myth does not serve as a symbolic framework for interpreting Randy Borman's role in Cofán society. For the Cofán, Borman is not an otherworldly divinity. "The Story of Maria" is more meaningful as a general expression of Cofán social structure in relation to nearly five hundred years of colonial and nation-state history. Its explication tells us how Cofán people survived their past and who they might become as they struggle to survive their present.

In all likelihood, the figure of Maria is related to the icons that Cofán obtained from Jesuit priests in the early Conquest era. Today, people remember the objects as "Little God" (statues) and "God's Skin" (paintings). In the story, Maria descends directly from the heavens, the home of Chiga (God or the sun).[1] Unbeknownst to the Cofán, the white-skinned and white-clothed Maria lives in a tree, away from the village. Cofán cosmology identifies tree trunks as the abodes of many supernatural agents, especially *cocoya* who bear physical and behavioral similarities to Cocama. Unlike most *cocoya*, however, Maria is a beneficial force. Her main duty is to protect the Cofán from disease and misfortune. Like a shaman, she fights off sickness and enemies from the social and spatial periphery of the village.

Maria's foreignness is reminiscent of a shaman's ethnic ambiguity. Nevertheless, she exhibits no moral ambivalence. In the myth, her main accomplishment is the robust health and steady growth of the Cofán population. People harness Maria's productive powers by building her a house. Spanish priests arrive and invert the situation. They force the Cofán to build a house into which the Cofán, rather than Maria, are lured. Subsequently, colonial soldiers massacre them. The differential history of violence explains a contemporary self-representation. Alluding to a more intense experience of decimation, the Aguarico Cofán describe themselves as quiet, subdued, and *opatssi*. In contrast, they assert that the Cofán of the San Miguel, many of whom live in Colombia, maintain their aggressive demeanor in the form of superior shamanic power.

After the massacres, the priests return. They steal Maria and install her in a Cocama house in Pasto, a city that Cofán people attacked in the colonial era. When a group of Cofán go to see her, she appears first as an anaconda.[2] Only after being blessed, baptized, and instructed by a priest can the Cofán see Maria in her familiar, beneficent form. Nevertheless, her protective services remain under the control of Cocama, whose power and population grow as the Cofán nation dwindles. To the Cofán who have not received the priest's blessing, Maria continues to appear as a *cocoya*, purely malevolent rather than purely benevolent.

"The Story of Maria" represents the division of epistemological labor in Cofán society. Maria's home corresponds to a shaman's *yaje* house. It also appears as a community school and a Cocama church. Both buildings contain mysterious writings. Only a few people can decipher the works, which foretell two catastrophes. In the first instance, children read the book in Maria's house-cum-school and prevent Chiga from taking her, thereby performing a shamanistic defensive action at one level of remove (i.e., they protect their protector). In the second instance, a literate Cofán reads the book in Maria's Pasto house. The esoteric knowledge tells only of death and the world's end. Perhaps Cofán people can purchase salvation at the price of conversion. The story, however, never poses or answers the question. Either way, Maria remains in a Cocama city. Never again will she return to ensure a life of health and plenty in Cofán territory.

"The Story of Maria" expresses more than four centuries of Cofán politics and sociality. After the pre-Conquest ferocity of the culture hero Erición and the early bellicosity of Cofán warriors who martyred priests and burned colonial towns, the Cofán did their best to avoid violence. Although Cofán people have always traveled to distant lands to trade, from the 1700s on, Cofán society appears to have revolved around the ideal of *opatssi* existence. The fundamental condition of the newly desired state was isolation from external power and violent difference. Only the mediating work of shamans and *na'su* could help them secure the life they sought.

In the new millennium, the great majority of Ecuadorian Cofán continue to orient their actions around the *opatssi* ideal. The collective pursuit of the valued condition produces a satisfying and meaningful existence in lowland communities. The style of life, however, is increasingly difficult to maintain in a world of massive colonization, environmental destruction, and Colombian warfare. The sociopolitical situation creates a contradiction: as threats become more intense, an *opatssi* life becomes more desirable, thereby creating a deep obstacle to the mobilization of the Cofán nation.

Randy Borman and other leaders are working to overcome the contradiction by convincing Cofán people of the necessity of collective action. They seek to reorient Cofán values toward the desirability of a warrior-like aggressiveness, which is a necessary shift if the Cofán are to overcome their contemporary enemies. The movement is an exercise in intentional social transformation. It is not, however, a production ex nihilo. Rather, it is a struggle to generalize a minority value orientation, which remains a latent possibility in the Cofán imaginary.

Borman's confidence in contemporary Cofán agency is the guiding force behind the Cofán experiment. Although the movement depends on transforming Cofán sensibilities, it also entails the integration of novel institutions into a lasting way of life. Ideally, the FSC will intensify its Quito operations. More Cofán people, as well as non-Cofán allies, will work to find funds, execute projects, and expand the Cofán nation's legalized territory. Additional Cofán villages will develop conservation systems that expand on the experience of Zábalo. With the aid of the FSC, communities will attract more researchers and management initiatives to their forests. The park guard force will continue to blanket the entirety of Cofán territory with monitoring and protection programs. Hopefully, tourists, too, will return. And if all goes well, the Education Project will continue to grow, producing new generations of skilled leaders. Land rights are still shaky; income levels are still insufficient; transnational alliances are still uncertain; educational successes are still rare; and far too much continues to depend on Borman, who is as fragile and mortal as any other human being. Against all odds, however, the balance of forces has tipped in favor of the Cofán.

Due to nation-state encroachment, world-systemic entanglement, and Cofán political action, the kind of "contact situation" (Turner 1988:239) portrayed in "The Story of Maria" no longer exists. Sporadic interaction, conquering armies, and the possibility of flight are things of the past. Certainly, similar dynamics are still in play: Cofán people formed Zábalo as an escape from petroleum-based development, and it continues to function as a refuge from Colombian violence. Moreover, people continue to view their leaders in terms of the distance and difference of shamanlike *na'su*. Nevertheless, the Cofán experiment is institutionalizing a novel structure of interaction between Cofán people and the actors and forces that encompass them.

It is impossible to predict whether the Cofán experiment will succeed in establishing a stable and productive relationship between Cofán society and a sociopolitical field composed of Ecuadorian neoliberalism; Colombian violence; U.S. foreign policy; global capitalism; and transnational networks of science, conservation, and human rights. Furthermore, it is impossible to know what effect such a development would have on the basic dynamics of Cofán life. I remain convinced that the value of *opa* continues to form the core of Cofán identity and society. It composes people's hopes and desires, and it orients their actions and satisfactions. The value, however, can only retain its affective power in a specific historical structure.

Although its connotations of calm, health, peace, equality, and good-natured hedonism differentiate it from a Nietzschean *ressentiment*, the value of *opa* is reactionary in nature. Its meaning is constituted against the violent and structured life of the shaman, who stands between the *opa a'i* and a cosmos composed of chaos and death. At its core, the value of *opa* is grounded in a flight from a "hostile external world" (Nietzsche 1992:473). Opposing such reactionary modes of valuation, Nietzsche argues for a proactive stance, which affirms the darkly expansive aspects of life:

> [This is] the noble mode of valuation: it acts and grows spontaneously, it seeks its opposite only so as to affirm itself more gratefully and triumphantly— its negative concept "low," "common," "bad" is only a subsequently-invented

pale, contrasting image in relation to its positive basic concept—filled with life and passion through and through. (Nietzsche 1992:473)

Nietzsche's conceptualization bears a similarity to the life metaphor of the jaguar. And it is not entirely dissimilar to Borman's desire for strong enemies and dramatic battles. Although it would be unwise to equate Nietzsche's vision of noble morality with either position, I do think that shifting the basic relationship between Cofán society and the realms that encompass it will produce substantial changes in Cofán sociality. The distinctive ethos that orients life in Cofán communities exists in opposition and reaction to centuries of sickness, violence, and fear. The problems continue to plague Cofán people in contemporary forms. In today's world, the old options of sociopolitical isolation and shamanic mediation are either insufficient or unavailable. A more direct form of collective confrontation is necessary, and it is now under way.

As a means, the struggle clashes with Cofán people's desire to lead *opatssi* lives. At an even deeper level, however, the end toward which Cofán people direct their mobilization—a secure and satisfying position in the contemporary world—is also problematic. Its realization will remove the conditions of antagonism, anxiety, and privation that make the dream of *opatssi* existence so seductive. In other words, if Cofán people produce their desired future, they will create a situation in which their central socioexistential value no longer provides a meaning and a telos to their lives. In the simplest terms, they will cease to be who they currently are. The Cofán must fight and win many battles, however, before the paradoxical dynamic takes on a social life. And by the time they do achieve their objectives—if they ever do—the contradiction may no longer appear as a problem to a nation that has remembered what it was like to take the offensive.

LESSONS AND POSSIBILITIES

Randy Borman, the people of Zábalo, and the greater Cofán nation are struggling to overcome a trying set of circumstances by creating a way of life that integrates indigenous culture, environmental conservation, scientific research, and an advantageous political-economic position. No one would have predicted their success forty, thirty, or even ten years ago. Despite repeated warnings of their demise, Cofán people continue to enjoy, to protect, and to profit from over 430,000 hectares of the world's most biologically diverse territory. Although their achievements are still provisional, their story provides reasons for hope when considering the fate of the world's cultural

Cofán territory on the Aguarico River, looking west to the Andes (Photo by Michael Cepek)

and biological diversity. In the next few pages, I offer a set of lessons that activists, advocates, and analysts can draw from Cofán experiences.

Indigenous Politics

With regard to struggles for indigenous rights, this book provides the empirical material for a number of useful insights. When considering the Cofán case, it is important to remember the dual senses of *representation*. First, the word refers to the ways in which scholars and other actors work to construct adequate portrayals of the inherently complex nature of peoples' social, cultural, and political experience. And second, it points to the means by which representatives speak and act for their constituents in local, national, and international contexts.

Throughout the book, I emphasized the nuanced and paradoxical nature of Cofán social logic and political perspectives. Although I am not afraid to make concrete claims about "Cofán culture," Cofán identity practices and political traditions are complicated, shifting, and contradictory. Supporters of indigenous struggles must be willing to accept that the objects of their aid do not fit outsiders' assumptions, either as groups or as individuals. Western commentators and advocates need to break free from their ideas of how ethnicity, race, and power function in the political projects of people with unique notions of themselves and the world.

The existence of complexity and contradiction, however, does not mean that there is no discernible structure to the political situations of contemporary indigenous peoples. Rather, it suggests that we can only understand the relevant dynamics through careful listening; a willingness to alter our assumptions; and long-term, critical, and collaborative investigation. Many anthropologists hold that realist accounts of indigenous culture and politics are inevitably essentializing and inherently problematic. In contrast, many Western advocates assume that they can quickly and confidently grasp the "true" nature of indigenous perspectives, which they construe in terms of their own preconceptions. The best option is to maintain a distance from both stances, working to discover the patterns and structures but constantly questioning one's biases and conclusions. Whenever possible, it is important to discuss our developing insights with the people about whom we write.

The issues become even more complicated when considering the objectives of indigenous struggles. There is often a disjuncture between the discourse of spokespeople and the viewpoints of their constituents. In the Cofán case, there is no consistent and universal political perspective. Although all Cofán people want to establish a secure position for themselves in the contemporary world, some enjoy the ease of community life while others yearn to engage the power and difference of external enemies. More than one value orients Cofán political action, and they work together in unexpected ways. Without exception, studying the relationship between values and politics is a difficult endeavor. If we forego it, however, we cannot understand what peoples are fighting for and how they are doing it—as well as how we might help them, if they desire our aid.

An additional observation that is essential to the study and support of indigenous struggles is the challenge of incorporating new institutional structures into ongoing political movements. For the Cofán, the NGO is both promising and painful. Working as an NGO demands negotiating new kinds of practical, social, and psychological strains. Moreover, it necessitates building partnerships with ethnic others. The work of non-Cofán collaborators is essential for mastering urban bureaucracies, alien social spaces, and Byzantine accounting procedures.

The Cofán case teaches us that many peoples have good reasons for making ethnic others into essential players in their collective organizations. Moreover, it suggests that observers should be extremely wary of accusing indigenous leaders of corruption, inauthenticity, or ineffectiveness. The work of running an NGO is a culturally complicated affair. It is essential to listen carefully to the explanations of indigenous officials as well as the critical commentary of the people they represent. There is always a backstory, and it often involves dynamics that clash with the assumptions of Western observers. A willingness to look past one's political and ethical instincts is key.

The Cofán case offers one more insight that is helpful for grasping the practice of indigenous politics. The ideology of Western democracy, as well as stereotypical representations of native peoples, suggest that indigenous struggles are fundamentally egalitarian endeavors. When considering Cofán accomplishments, however, it is clear that one or more exceptional individuals can account for a great deal of the content and consequences of political projects. Most non-Western or small-scale societies do not practice what Marshall Sahlins calls "heroic history" (Sahlins 1985:108), and few Amazonian cultural logics merit the label "messianic" (Cepek 2009; Veber 2003). Nevertheless, the Cofán case demonstrates how important one individual can be, even if they must actualize their vision through deep involvement with multiple social fields. In other words, just because an indigenous project exhibits an uneven division of labor, it does not mean that the effort is any less important or legitimate because of it.

Local Peoples and Environmental Conservation

Insights from the Cofán case can help to inform the construction of appropriate approaches to environmental conservation in the global south. First and foremost, the Cofán experiment demonstrates that a "protectionist," "people-free," or "fortress conservation" strategy will not produce desirable results in many regions. There are good reasons to collaborate with such peoples as the Cofán. Like most indigenous nations, however, the Cofán do not conform to the myth of the "ecologically noble savage" (Redford 1991). They do not view the forest or its denizens as "sacred," and Cofán cosmology contains few, if any, conservationist lessons. Even though it articulates with long-standing elements of Cofán culture and social structure, Cofán conservationism emerged in a historical, political, and economic conjuncture. Only recently did Cofán people learn to value the forest as an essential foundation of their way of life. Their decision to base their political future on conservationist commitments, moreover, is an even more recent development.

After witnessing the large-scale destruction of their territory at the hands of oil companies and colonists, Cofán people decided to restructure their environmental relations to ensure their continuity as a people. Their conservationist turn, however, did not make them automatic allies of Western environmentalists. After nearly five hundred years of oppression, they do not trust the goodwill, idealism, or altruism of any Gringo or Cocama. Surely, many indigenous peoples feel the same. When northerners show concern for their lands and lives, the Cofán assume that the outsiders are acting in their own interest. Conservationists must realize that the reaction is justified. It is even more sensible for such peoples as the Cofán, who do not doubt their own ability to know and to care for their forests. The perspective is compounded by the alien nature of Western conservation's logics, techniques, and instruments, which are beyond the means and familiarity of many subaltern peoples.

Cofán people desire long-term partnerships rather than hit-and-run projects. Most outsiders make promises that they do not keep, and their commitments are temporary at best. Years of return visits allow for trust and reciprocity. For people in Cofán communities, simply receiving copies of research reports and journal articles is important, even if the publications are in languages they do not understand. Additionally, learning people's names, exchanging gifts with them, eating in their homes, and mastering a bit of their language are deeply appreciated gestures. Indigenous peoples have witnessed so many "helpful" outsiders come and go that showing a real commitment to understanding their situations can lay the groundwork for more effective collaboration.

Although following social protocol is important, the Cofán desire more substantial political-economic reciprocity. They want outsiders to help them achieve legal control over their traditional territory. They need opportunities for modest but sustained income, whether as scientific researchers, project planners, park guards, or logistical coordinators for conservation efforts. The work builds on their impressive skills in the forest environment, and it allows them to reside in their homeland. Cofán people realize that they are protecting the Amazonian environment for the whole world. They believe that they deserve payment for their work, especially when it obeys outside logics and benefits outside actors.

Although some Western environmentalists might respond that there is not enough money to support permanent employment for "indigenous conservation professionals," no one has done the calculations. Cofán conservationism provides income to hundreds of people, and it ensures the effective protection of more than one million acres of forestland. Currently, the entire enterprise—including the support of FSC employees in Quito as well as the education of Cofán students—costs approximately one million dollars a year. After watching dozens of Western NGOs attempt dozens of ineffective projects over dozens of years, Cofán people are skeptical of any argument based on the notion of economic feasibility. They know that some Western conservation workers make six-figure salaries. Is it really so absurd, they ask, to find a more efficient way to spend the money? From their perspective, the situation is senseless. They believe that their successes are important innovations in a world of failed conservation initiatives, especially in the global south. Payment is an essential element of the system they have created, and they want outsiders to understand why.

Most environmentalist NGOs are hesitant to fund projects that are not explicitly "ecological" in nature. The Cofán case suggests that they should expand their perspectives. For years, Cofán leaders have sought funds to educate promising students in high-quality institutions. Randy Borman has been essential to the success of Cofán conservationism. Without future leaders who can take over his work, Cofán accomplishments will be short-lived. If conservationists want their interventions to become part of a long-term political reality, they must be willing to think outside the box.

Supporting indigenous peoples' struggles to create globally capable leaders is an important aspect of the fight to protect the world's threatened ecosystems. The payoffs are seldom direct or assured. Some prospective leaders, for example, will decide to pursue other political or economic ventures after they receive their degrees. One thing is certain, however. If indigenous representatives are not capable of negotiating national and transnational networks on their own terms, Western environmentalists will lose a promising set of potential allies. Without the concerted, informed, and educated efforts of Cofán representatives, northeastern Ecuador's natural landscapes would be doomed to destruction. Similar scenarios exist across the world.

Finally, Western conservationists who seek to alter local environmental practice need a good dose of humility. It is impossible to engineer culture and society. The Cofán case makes clear that short-term training and education programs rarely achieve the desired result of shifting local consciousness toward environmentalist ends. Even relatively intense and long-term efforts, such as the Field Museum's work in Zábalo, transport knowledge, technology, and skills in unanticipated ways. I am not arguing that all Western efforts to support local conservation are failures. I am, however, suggesting that we must rethink our paradigms and take the Cofán lead in terms of imagining the power we wield and the partnerships we construct.

Connected to the last point, I am skeptical of most arguments that describe conservation interventions as exercises in what Michel Foucault calls "governmentality" (1991). I do not deny that Western practitioners intend to remake indigenous subjectivities, or modes of value, belief, identity, experience, and desire. Indeed, they deploy a vast array of knowledge and technology in hopes of transforming local socioecological relations. Everything I have witnessed with the Cofán, however, convinces me that the efforts typically fail to change anything at all—especially the deep logic of indigenous ways of thinking, speaking, acting, and being.

The governmentality paradigm is a useful tool for studying policy and intent. It does not, however, capture the dynamics by which local discourse and practice inflect and deflect the objectives of outside interventions. As I argue in a recent article (Cepek 2011), the approach leads scholars to underestimate the critical, culturally framed consciousness of intervention's objects. Even more importantly, it construes efforts to form coalitional movements as little more than insidious power plays. In short, whereas the Cofán case suggests ways in which indigenous peoples, Western advocates, and academic analysts might engage in deep discussion and long-term collaboration toward the goal of effective conservation, thinking in terms of governmentality stymies action. More often than not, scholars committed to this paradigm hesitate to engage at all, they portray intercultural projects in an excessively negative light, and they deny local people's ability to rethink and retool the "aid" that comes their way. In short, their work cloaks projects for socioecological change in pessimism and paranoia. Their analytic blinders and political hypochondria provide little guidance or motivation to anyone invested in the future of the world's cultural and biological diversity. Theoretically, we can do a better job.

The Possibilities of Ethnography

Ethnographers can play important roles in projects for socioecological change. In Amazonia, they have a long history of activism and advocacy. Bruce Albert argues that the predominant mode of fieldwork is no longer "participant observation" but "observant participation" (1997). Anthropologists have contributed to indigenous struggles in many ways: documenting ancestral claims to territory, securing health and education services for local communities, obtaining funding for indigenous projects and organizations, publicizing rights violations, establishing forums for the airing of grievances and demands, lobbying corporate and state officials, creating cultural and historical documentation programs, mediating between indigenous organizations and Western NGOs, and helping to formulate national and international legal structures that recognize indigenous rights.

The Amazonianist Robin Wright recognizes the concrete contributions that anthropologists make to struggles for a diverse world. Nevertheless, he suggests that one of anthropology's most important missions is to disrupt debilitating notions of "political realism" (1988:382). Although I have engaged in direct collaborations with Cofán people, I wrote *A Future for Amazonia* as a response to Wright's call for a more "conceptual" intervention. I want the book to break open the sense of political and analytic closure that constrains our ability to think and work toward alternative futures. No one writing in the 1960s would have foreseen the twists and turns of the last forty years of Cofán history. At the time, observers identified the Cofán plight as one more chapter in a universal story of cultural and biological loss. But the Cofán proved them wrong. Their actions give us critical distance from a powerful narrative that often goes unquestioned.

Ultimately, *A Future for Amazonia* is about a utopian moment in global history. The world it engages, however, is actual rather than potential. Anthropology can help us conceptualize alternative futures by exploring existing liberatory projects. It uses empirical investigation rather than philosophical introspection to stimulate our imagination and ignite our hope. A different ethnographer might have written a very different account of the Cofán experiment. Nevertheless, I believe that my optimism is a reflection of real Cofán achievements rather than analytic bias or political daydreams. Either way, if this book has persuaded readers to reconsider the fate of the world's cultural and biological diversity—as well as their own contributions to struggles for indigenous rights and environmental conservation—then I will have achieved my goal.

Notes

INTRODUCTION

1. In this book, I use the A'ingae spelling system created by Bub Borman, Randy's father. The system is similar to that for Spanish with a few major exceptions: the use of double consonants to signal aspiration (e.g., *naccu* [residential group]); the use of an "m" or "n" after a vowel to signal nasalization (e.g., *anttia* [kin]); and the use of an apostrophe to signal a glottal stop (e.g., *a'i* [human being]). In addition, the vowel "u" is pronounced with unrounded labialization.

2. I have not been able to locate an original copy of the Friede article, so I make use of a version that Randy Borman's father keyed in, word for word, in an English translation.

3. To the best of my knowledge, their action was the first successful expulsion of an oil company from an indigenous South American community.

4. I am thinking here of ideas such as the "division of linguistic labor," as formulated by Hilary Putnam (1975).

5. At the time, Zábalo had not yet acquired legal personhood as a *centro*, and was thus identified in the early agreement simply as *la comunidad cofán de Zábalo* (the Cofán community of Zábalo).

CHAPTER 2

1. This version comes from the episode "American Son," which aired on CBS's *America Tonight* on June 8, 1994.

2. The relation between Cofán people and highland indigenous people is complicated. In the distant past, there were clear social relations between the groups. In more recent times, Cofán have had experiences with highland indigenous people as either immigrants to the lowlands, and thus as colonists, or through the national indigenous movement. Cofán people usually refer to highland indigenous people as Ccotacco'su Napo (Mountain Napo Runa), which is probably a recognition that both peoples speak Quichua, which Cofán refer to as Napongae.

3. The plural suffix *-ndeccu* is used only for human collectivities.

4. Randy Borman (personal communication) suggests that the use of the term "Cocama" derives from historically shifting reference patterns. In all likelihood, the term first applied to the indigenous Cocama-Cocamilla people who inhabited the area surrounding the confluence of the Napo and Marañón Rivers, with whom Cofán have had hundreds of years of trading relations. As mestizo people began to outnumber indigenous residents of the area, however, Cofán people shifted the term

to refer to them. Now, it is a term for the majority of Ecuador's citizens, as well as those of other Latin American countries.

5. Chapter 4 provides a detailed analysis of this term.

6. The A'ingae verb is *jeñañe*, the causative form of *jeñe* (to make sound). Cofán people use the lexeme to refer to the starting and handling of machines, such as airplanes, cars, outboard motors, and chain saws. The origin points to the noise of combustion. Now, the verb has the sense of driving, directing, and managing. When the man calls Borman the *jeña'su*, or "driver," of the community, he is making a metaphoric extension of the sense of driving a mechanized vehicle.

CHAPTER 3

1. "Pablo" is a pseudonym, as I want to protect the privacy of my consultant while discussing his moral ambiguity.

2. My understanding of "value" derives from a close study of the work of a number of anthropologists, sociologists, and philosophers (Bateson 1958; Benedict 1989; Bourdieu 1977; Geertz 1973; Graeber 2001; Joas 2000; Kockelman 2010; Marx 1977; Munn 1986; Parsons and Shils 1994; Taylor 1985; Turner 1984, 2003a). I provide a more detailed discussion of my approach in other publications (Cepek 2006, 2008a).

3. Like many other indigenous Amazonians, Cofán people care for the infants of adult wild animals that they have killed and eaten.

4. The Quichua words *mama* and *yaya* are gradually replacing the A'ingae words *chan* and *quitsa*. Nevertheless, both pairs are still in use.

5. I know of at least one case in which a young girl, who lives with her mother's parents, addresses her father as *du'su* (son-in-law), copying her grandparents' usage.

6. I use the masculine pronoun because men dominate in this realm of Cofán society, although interested women, too, can acquire power and become *curaga*.

7. The word *na'su* has many different meanings, depending on context: chief (of a community), leader (of a social group), owner (of an object), initiator (of a collective action), witness (of an event), and producer (of an utterance, artifact, or trace [e.g., footprint]). Historically, Cofán people also referred to community chiefs as *curaga*, *coen* (grown person), and *quini'cco'pa* (possessor of a staff of authority, which colonial and republican officials gave to indigenous intermediaries).

8. Cofán people elect community and federation officers in both community and intercommunity *bo'cho* (gatherings), where all adult men and women select leaders through open, verbal voting. They refer to the process as *na'suma somboeñe* (making leaders emerge).

9. Borman's proposal was successful, and the park guard force has been operational since 2003. I describe it in Chapters 4 and 5 and the conclusion.

CHAPTER 4

1. Both the FSC and the CSF have a board of directors, which include Borman and Cofán and non-Cofán allies. The boards, however, play little part in directing the activities of either organization.

2. When I visited in 2010, ICCA was inoperative because of a lack of funds. With new grants or payment for training courses, it will hopefully reopen its doors.

3. During 2001 and 2002, the monthly salaries of the employees were as follows: $500 for Borman, $400 for Espinosa and Aguinda, $300 for Lopez, and $200 for Quenamá. The numbers were well above Ecuador's minimum wage but well below the salaries at other Ecuadorian NGOs. It is my understanding that the salaries increased substantially after 2002. Nevertheless, funding crises continue to complicate the situation, as all the employees can go months without receiving a paycheck when one grant ends and another has yet to begin.

4. After the FSC purchased additional land, Borman's family moved into a house next to the FSC office.

5. Guayaquil is Ecuador's largest city, on the Pacific coast.

6. I wish to make clear that both Lopez and Borman gave me full access to their account books. In addition, they were willing to explain any confusion about financial matters. After working closely with them for years, I have no doubts regarding their ethical and professional standards. In other words, I saw absolutely no evidence of intentional misappropriation of funds at the FSC.

7. "CAIMAN" stands for "Conservation in Managed Indigenous Areas." Although it is not truly an organization, FSC employees understand it as such. Technically, CAIMAN is a Plan Colombia–related program of USAID. Chemonics International Inc., a U.S. consulting firm, is its main administrator. Chemonics hired a number of Ecuadorian and foreign individuals to work on CAIMAN, giving the program the appearance of an NGO. The consultants, rather than permanent Chemonics staff, are the main individuals who interface with the FSC about projects supported through the program.

8. Traveling from Zábalo to Lago Agrio requires an outboard motor, a good canoe, and hundreds of dollars for gasoline and oil, as well as wear and tear on the motor.

9. I describe community conservation practices in Chapter 5.

CHAPTER 5

1. Cofán elders' descriptions of the practice always include a means of determining whether the animal was "truly called" or simply encountered with no shamanic mediation. If a tapir were truly called, for example, a person would butcher it and find the seeds of the *yaje* admixture *opirito* (*Psychotria pschotriaefolia*) in its stomach.

2. In precolonial times, when thousands of indigenous people lived along the upper Aguarico in large settlements, subsistence must have been structured in a more systematic way. Hundreds of years and many demographic catastrophes later, however, Cofán people cannot imagine what the structure was.

3. Many Cofán made periodic use of Zábalo territory well before the founding of the community. For generations, it was a favored area for hunting and turtle egg–collecting trips. Generations earlier, Cofán and Siona people lived in mixed settlements both above and below Zábalo.

4. Some of the species, such as the raptors, are traditionally considered inedible, while others, such as caimans and macaws, were commonly eaten before the creation of the prohibitions.

5. An example is the prohibition on eating piping guans during adolescence. People fear that the white feathers of the birds will cause prematurely gray hair. The belief explains the Cofán nickname for Borman—Coyovi Ccashe (Old Piping Guan)—which he received because of his desire to hunt and eat piping guans during his youth, which explains his white hair today.

6. Even in the phrase "That animal is unkillable," my informant distances the maxim from the doing of the president by using the *te* reportative particle. A more accurate translation would be, "It is said that that animal is unkillable," with the community implicitly identified as the agent.

7. "*Tsesuma tsonja*" (Do that), with the *-ja* command suffix on the verb *tsoñe* (to do).

8. To the best of my knowledge, no one has ever committed the same infraction more than once.

9. The word is *pansha'ombi*, which is the negative adjectival form of the verb *panshañe* (to pass or exceed).

10. Shortly after I defended my dissertation in 2006, the ECP joined the Field Museum's Center for Cultural Understanding and Change (CCUC) under the newly created Division of Environment, Culture, and Conservation (ECCo). A few years later, the CCUC merged with the ECP. ECCo is now the official designation for the museum branch that continues the work of the ECP. I retain the former label in this book because it was the only term used by both Cofán people and museum personnel during my fieldwork. In addition, many museum employees continue to use the label to identify themselves and their work in unofficial contexts. I began cooperating with the ECP as a volunteer in 2001. I worked part-time for them between February of 2004 and April of 2006, and I continue to serve as an unpaid ECCo fellow. I do not believe that my relationship with the ECP compromises the objectivity of my account. The ECP is most interested in my critical commentary on their efforts, and ECP personnel have absolutely no editorial oversight of the book. Nevertheless, I wish to state that I consider myself to be a conservationist, and that I admire the expertise, openness, motivation, and ethical approach of the ECP.

11. The coordinators were responsible for directing the work and coordinating data entry. In 2001 and 2002, they received $150/month. Monitors received $100/month. The ECP initially understood payment as a means of sparking community interest in a foreign set of practices. Gradually, they became convinced that it was essential to Cofán participation, as I describe below.

12. According to most narratives that I recorded, the turtle project began when Eduardo Yiyoguaje collected a nest of hatchlings and kept them in a bucket next to his house. Keeping pets is a common practice in Zábalo, and there was nothing extraordinary about his action. By coincidence, visiting tourists saw the turtles and gave a small sum of money to Borman to help the community care for them. After a few years of experimentation, including the building of small ponds, outside scientists and NGOs became interested in studying and aiding Cofán experiments in raising turtles. The outsiders donated materials for making bigger pools. Slowly, the entire community became involved by collecting the hatchlings and putting them into the pools. Only with the entrance of the Field Museum, however, did monitors and coordinators begin working on the project and keeping systematic data on its progress.

13. As a class of activities, *semañe* does not cover hunting and fishing.

14. Although it might seem paternalistic to prohibit workers from carrying guns while doing their census work, the rule does make sense. For example, while walking with a monitor, we noticed a tinamou on the ground. The worker stopped to record the sighting, holding his machete at his side. Suddenly, another tinamou took flight nearby. The worker instinctively flung his machete at the second bird. He missed, but his uncontrollable reaction visibly embarrassed him. He knew that it was prohibited to attempt to kill anything while doing project work.

15. The suggested change never occurred. Project workers argued that they had developed a deep knowledge of project activities that other community members did not share.

CHAPTER 6

1. I introduced Antonio at the beginning of Chapter 3.

2. I provide a detailed description of Cofán marriage practices as well as all rituals associated with the life cycle in my doctoral thesis (Cepek 2006).

3. Since 2002, when I left the field, I believe that the annual costs have risen substantially.

4. Within the last few years, the MacArthur Foundation began paying for a large portion of the Education Project's costs.

5. I suggested to the teachers that the Cofán excelled in physical activities because they were two or more years older than their classmates. The teachers disagreed with my logic, explaining that the Cofán have superior "natural agility."

6. Hugo Lucitante reported that his year at the Alliance Academy was more academically rigorous than his time at schools in Seattle.

7. Borman has paid for the majority of his children's education with his own money. His parents have helped, too. It was impossible for me to track the minute-by-minute expenses of the FSC, but I do think that some of the FSC funds were used for Alliance Academy fees.

8. While I was in Zábalo from July of 2001 to June of 2002, Hugo Lucitante spent a year at the QCC and the Alliance Academy.

9. Quenamá also fulfills the role, but she is less of an authority figure than Borman.

10. By 2009, two Cofán girls had become part of the Education Project.

11. Bolivar Lucitante, Hugo's father, has a Cocama father and kin who live close to the city of Coca. After I left the field, he sent his daughter to live with them so that she could learn Spanish in a Cocama school.

12. Amelia chooses to use "Quenamá" as her last name even though her father's last name is "Yiyoguaje." Her maternal grandmother and grandfather (Aniseto Quenamá) raised her, and she wants her public identity to reflect her history.

13. Bolivar has served as president and vice president of Zábalo multiple times. The Spanish he learned in Ecuador's army, as well as his general affability, make him a favorite of tourists and visiting researchers.

14. I changed the name of this leader to protect his privacy.

CONCLUSION

1. I have not discussed the meaning of Chiga in this book. The figure now corresponds to a Cofán version of the Christian God, although the word also means "sun" in A'ingae. Contemporary Cofán who consider themselves to be Christian are called *Chigave in'jan'cho* (believers in God). The label, however, is flexible. Many people say that a *Chigave in'jan'cho* is anyone who does not get drunk, does not cheat on their spouse, does not fight, and does not kill. Shamanic discourse and Cofán cosmology make little mention of Chiga, although there are a few myths in which he figures (Borman 1990). Cofán people say that Chiga used to live on Earth. They claim that they knew him before the Spanish came, as described in "The Story of Maria."

2. The anaconda is a clearly malevolent figure in Cofán cosmology. It often appears as an insatiable devourer of humans. Its only positive function derives from its position as the *na'su* of fish, which it can exchange with a shaman for a meal of human flesh. A powerful shaman, however, can make a deer or a paca "appear" as a human, which he feeds to the anaconda to satisfy its hunger. Interestingly, the anaconda is the one nonhuman figure that combines the acquisitive and defensive functions of the shaman. Perhaps the double meaning explains its appearance in Pasto. More than the origin of Cocama violence, that city is also the source of manufactured goods. In the myth, however, there is no mention of Maria's role in the acquisition of desired commodities. She functions purely to reproduce the internal basis of Cofán society.

Bibliography

ADELMAN, JEREMY

2002. "Andean Impasses." *New Left Review* 18:40–72.

AGRAWAL, ARUN

2005a. "Environmentality: Community, Intimate Government, and Environmental Subjects in Kumaon, India." *Current Anthropology* 46(2):161–190.

2005b. *Environmentality: Technologies of Government and the Making of Subjects.* Durham: Duke University Press.

ALBERT, BRUCE

1997. "'Ethnographic Situation' and Ethnic Movements: Notes on Post-Malinowskian Fieldwork." *Critique of Anthropology* 17(1):53–65.

BARTH, FREDERIK

1969. Introduction. *Ethnic Groups and Boundaries: The Organization of Cultural Difference*, ed. F. Barth. Boston: Little, Brown.

BATESON, GREGORY

1958. *Naven: A Survey of the Social Problems Suggested by a Composite Picture of the Culture of a New Guinea Tribe Drawn from Three Points of View.* Palo Alto: Stanford University Press.

BENEDICT, RUTH

1989. *Patterns of Culture.* Boston: Houghton Mifflin.

BERGER, PETER L., AND THOMAS LUCKMANN

1966. *The Social Construction of Reality: A Treatise in the Sociology of Knowledge.* Harmondsworth, UK: Penguin.

BERLINGER, JOE, DIR.

2009. *Crude.* First Run Features.

BHASKAR, ROY

1989. *Reclaiming Reality: A Critical Introduction to Contemporary Philosophy.* London: Verso.

1998. *The Possibility of Naturalism: A Philosophical Critique of the Contemporary Human Sciences.* London: Routledge.

BLOCH, ERNST

1986. *The Principle of Hope.* 3 vols. Cambridge: MIT Press.

BODLEY, JOHN

1972. "A Transformative Movement among the Campa of Eastern Peru." *Anthropos* 67:220–228.

BORMAN, MARLYTTE B.

    1990. *Cofán History and Cosmology as Revealed in Their Legends*. Quito: Instituto Lingüístico de Verano.

BORMAN, RANDALL B.

    1996. "Survival in a Hostile World: Culture Change and Missionary Influence among the Cofán People of Ecuador, 1954–1994." *Missiology* 24(2):185–200.

    1999. "Cofán: Story of the Forest People and the Outsiders." *Cultural Survival Quarterly* 23(3): 40–50.

    2002. "Políticas cofanes de conservación, manejo y uso de recursos biológicos." Unpublished manuscript.

    2009. "A History of the Río Cofanes Territory." In *Rapid Inventory No. 21: Ecuador: Cabeceres Cofanes-Chingual*, ed. C. Vriesendorp, William S. Alverson, Álvaro del Campo, Douglas F. Stotz, Debra K. Moskovits, Segundo Fuentes Cáceres, Byron Coronel Tapia, and Elizabeth P. Anderson, 222–227. Chicago: Field Museum of Natural History.

BOURDIEU, PIERRE

    1977. *Outline of a Theory of Practice*. Trans. R. Nice. Cambridge: Cambridge University Press.

BRAUN, BRUCE

    2000. "Producing Vertical Territory: Geology and Governmentality in Late Victorian Canada." *Cultural Geographies* 7(1):7–46.

    2003. *The Intemperate Rainforest*. Minneapolis: University of Minnesota Press.

BREMMER, IAN

    2006. "The Little Guys of the Oil Business: Meet the 'Marginal Producers.'" *Slate*, May 10. http://www.slate.com/articles/news_and_politics/the_gist/2006/05/the_little_guys_of_the_oil_business.html.

BROSIUS, PETER

    1997. "Endangered Forest, Endangered People: Environmentalist Representations of Indigenous Knowledge." *Human Ecology* 25(1):47–69.

    1999a. "Analyses and Interventions: Anthropological Engagements with Environmentalism." *Current Anthropology* 40(3):277–309.

    1999b. "Green Dots, Pink Hearts: Displacing Politics from the Malaysian Rainforest." *American Anthropologist* 101(1):36–57.

BROWN, MICHAEL F.

    1991. "Beyond Resistance: A Comparative Study of Utopian Renewal in Amazonia." *Ethnohistory* 38(4):388–413.

    1993. "Facing the State, Facing the World: Amazonia's Native Leaders and the New Politics of Identity." *L'Homme* 33(2–4):307–326.

BROWN, MICHAEL F., AND EDUARDO FERNÁNDEZ

    1995. *War of Shadows: The Struggle for Utopia in the Peruvian Amazon*. Berkeley: University of California Press.

BRYSK, ALISON

    1996. "Turning Weakness into Strength: The Internationalization of Indian Rights." *Latin American Perspectives* 23(2):38–57.

    2000. *From Tribal Village to Global Village: Indian Rights and International Relations in Latin America*. Stanford: Stanford University Press.

BUEGE, DOUGLAS J.

    1996. "The Ecologically Noble Savage Revisited." *Environmental Ethics* 18:71–88.

CABODEVILLA, MIGUEL ANGEL

    1996. *Coca: La región y su historia.* Pompeya, Ecuador: CICAME.

    1997. *La selva de los fantasmas errantes.* Pompeya, Ecuador: CICAME.

CARNEIRO DA CUNHA, MANUELA

    1973. "Logic of Myth and Action: The Canela Messianic Movement of 1963." Unpublished manuscript.

    1998. "Points of View on the Amazonian Forest: Shamanism and Translation." *Mana* 4(1):7–22.

CARNEIRO DA CUNHA, MANUELA, AND MAURO W. BARBOSA DE ALMEIDA

    2000. "Indigenous People, Traditional People, and Conservation in the Amazon." *Daedelus* 129:315–338.

CASTORIADIS, CORNELIUS

    1998. *The Imaginary Institution of Society.* Trans. K. Blamey. Cambridge: MIT Press.

CENTRO PARA DERECHOS ECONÓMICOS Y SOCIALES

    1994. *Violaciones de derechos en la Amazonia ecuatoriana: Las consequencias humanas del desarollo petrolero.* Quito: Abya-Yala.

CEPEK, MICHAEL L.

    1996. "Reorganization and Resistance: Petroleum, Conservation, and Cofán Transformations." Bachelor's thesis, Anthropology, University of Illinois at Urbana-Champaign.

    2006. "The Cofán Experiment: Expanding an Indigenous Amazonian World." Ph.D. diss., Department of Anthropology, University of Chicago.

    2008a. "Bold Jaguars and Unsuspecting Monkeys: The Value of Fearlessness in Cofán Politics." *Journal of the Royal Anthropological Institute* 14:331–349.

    2008b. "Essential Commitments: Identity and the Politics of Cofán Conservation." *Journal of Latin American and Caribbean Anthropology* 13(1): 196–222.

    2009. "The Myth of the Gringo Chief: Amazonian Messiahs and the Power of Immediacy." *Identities: Global Studies in Culture and Power* 16(2):227–248.

    2011. "Foucault in the Forest: Questioning Environmentality in Amazonia." *American Ethnologist* 38(3):501–515.

CERÓN, CARLOS E.

    1995. *Etnobiología de los cofanes de Dureno.* Quito: Museo Ecuatoriano de Ciencias Naturales.

CHAPIN, MAC

    2004. "A Challenge to Conservationists." *World Watch Magazine* (November/December):17–31.

CLASTRES, HÉLENE

    1978. *Terra Sem Mal: O Profetismo Tupi-Guaraní.* São Paulo: Brasiliense.

COLLINS, JENNIFER

    2004. "Linking Movement and Electoral Politics: Ecuador's Indigenous Movement and the Rise of Pachakutik." In *Politics in the Andes: Identity,*

Conflict, Reform, ed. J.-M. Burt and P. Mauceri, 38–57. Pittsburgh: University of Pittsburgh Press.

CONKLIN, BETH

1997. "Body Paint, Feathers, and VCRs: Aesthetics and Authenticity in Amazonian Activism." *American Ethnologist* 24(4):711–737.

2002. "Shamans versus Pirates in the Amazonian Treasure Chest." *American Anthropologist* 104(4):1050–1061.

CONKLIN, BETH, AND LAURA R. GRAHAM

1995. "The Shifting Middle Ground: Amazonian Indians and Eco-Politics." *American Anthropologist* 97(4):695–710.

CROCKER, WILLIAM

1967. "The Canela Messianic Movement: An Introduction." *Atas do Simpósio sobre a Biota Amazonica* 2:69–83.

DARIER, ERIC

1999. "Foucault and the Environment: An Introduction." In *Discourses of the Environment*, ed. E. Darier, 1–34. Oxford: Blackwell.

DEWEY, JOHN

1929. *Experience and Nature*. Chicago: Open Court.

EDELI, DAVID, AND ZACHARY HURWITZ

2003. "There Can Be No Peace without Indians at the Table: A Narrative from Armando Valbuena." *Cultural Survival Quarterly* (Winter):13–17.

EDELI, DAVID, AND KYLE RICHARDSON

2003. "Colombia's Expanding War." *Cultural Survival Quarterly* (Winter): 63–69.

EFC NEWS SERVICES

2011. "Ecuador: Over 50 Pct of Oil Exports Went to China in September." http://www.menafn.com/qn_news_story.asp?storyid=%7B0b6cd0e8-3d41 -45d3-8ea3-bb669f8d66e2%7D

ESCOBAR, ARTURO

1994. *Encountering Development: The Making and Unmaking of the Third World*. Princeton: Princeton University Press.

FERGUSON, JAMES

1994. *The Anti-Politics Machine: "Development," Depoliticization, and Bureaucratic Power in Lesotho*. Minneapolis: University of Minnesota Press.

FIELD MUSEUM

2005. Environmental and Conservation Programs. Accessed March 1. http://www.fieldmuseum.org/research_collections/ecp.

FISCHER, RAFAEL

2007. "Clause Linkage in Cofán (A'ingae), a Language of the Ecuadorian-Colombian Border Region." In *Language Endangerment and Endangered Languages: Linguistic and Anthropological Studies with Special Emphasis on the Languages and Cultures of the Andean-Amazonian Border Area*, ed. L. Wetzels, 381–400. Leiden: CNWS Publications.

FISHER, WILLIAM

1997. "Doing Good? The Politics and Antipolitics of NGO Practices." *Annual Review of Anthropology* 26 (October):439–464.

FLETCHER, NATALY

2003. "Advocates or Obstacles?: NGOs and Plan Colombia." *Cultural Survival Quarterly* (Winter):18–20.

FOUCAULT, MICHEL

1991. "Governmentality." In *The Foucault Effect: Studies in Governmentality*, ed. G. Burchell, C. Gordon, and P. Miller, 87–104. Chicago: University of Chicago Press.

FRIEDE, JUAN

1952. "Los kofán: Una tribu de la alta Amazonia colombiana." XXX International Congress of Americanists.

FRIEDMAN, JONATHAN

1996. "The Politics of De-Authentification: Escaping from Identity, A Response to 'Beyond Authenticity' by Mark Rogers." *Identities: Global Studies in Culture and Power* 3(1–2):127–136.

FUNDACIÓN ZIO-A'I

2000. *Plan de vida del pueblo cofán y cabildos indígenas del Valle del Guamuéz y San Miguel.* Bogotá: Fundación Zio-A'i, Unión de Sabiduría.

GEERTZ, CLIFFORD

1973. *The Interpretation of Cultures.* New York: Basic Books.

GRAEBER, DAVID

2001. *Toward an Anthropological Theory of Value: The False Coin of Our Own Dreams.* London: Palgrave.

2004. *Fragments of an Anarchist Anthropology.* Chicago: Prickly Paradigm Press.

GRAHAM, LAURA R.

2002. "How Should an Indian Speak? Amazonian Indians and the Symbolic Politics of Language in the Global Public Sphere." In *Indigenous Movements, Self-Representation, and the State in Latin America*, ed. K. Warren and J. Jackson, 181–228. Austin: University of Texas Press.

2005. "Image and Instrumentality in a Xavante Politics of Existential Recognition: The Public Outreach Work of Eténhiritipa Pimentel Barbosa." *American Ethnologist* 32(4):622–641.

GREENE, SHANE

2009. *Customizing Indigeneity: Paths to a Visionary Politics in Peru.* Palo Alto: Stanford University Press.

GUSTAFSON, BRET

2009. *New Languages of the State: Indigenous Resurgence and the Politics of Knowledge in Bolivia.* Durham: Duke University Press.

HALE, CHARLES R.

2006. "Activist Research v. Cultural Critique: Indigenous Land Rights and the Contradictions of Politically Engaged Anthropology." *Cultural Anthropology* 21(1):96–120.

HARVEY, DAVID

2000. *Spaces of Hope.* Berkeley: University of California Press.

HEIDEGGER, MARTIN
 1996. *Being and Time.* Trans. J. Stambaugh. Albany: State University of New York Press.

HOOPER, JOSEPH
 1991. "The Gringo Chief." In *Insight Guide to Ecuador,* ed. P. Barret, 99–124. Maspeth, NY: Langenscheidt Publishers.

HUGH-JONES, STEPHEN
 1996. "Shamans, Prophets, Priests and Pastors." In *Shamanism, History, and the State,* ed. N. Thomas and C. Humphrey, 32–75. Ann Arbor: University of Michigan Press.

HURTIG, ANNA-KARIN, AND MIGUEL SAN SEBASTIÁN
 2002. "Geographical Differences in Cancer Incidence in the Amazon Basin of Ecuador in Relation to Residence Near Oil Fields." *International Journal of Epidemiology* 31:1021–1027.
 2004. "Incidence of Childhood Leukemia and Oil Exploitation in the Amazon Basin of Ecuador." *International Journal of Occupational Environmental Health* 10(3):245–250.

HYLTON, FORREST
 2003. "An Evil Hour: Uribe's Colombia in Historical Perspective." *New Left Review* 23:51–93.

JACKSON, JEAN E.
 1995. "Culture, Genuine and Spurious: The Politics of Indianness in the Vaupés, Colombia." *American Ethnologist* 22(1):3–27.

JACKSON, JEAN E., AND KAY WARREN
 2005. "Indigenous Movements in Latin America, 1992–2004: Controversies, Ironies, New Directions." *Annual Review of Anthropology* 34:549–573.

JOAS, HANS
 1996. *The Creativity of Action.* Trans. J. Gaines and P. Keast. Chicago: University of Chicago Press.
 2000. *The Genesis of Values.* Chicago: University of Chicago Press.

KANE, JOE
 1995. *Savages.* New York: Knopf.

KIMERLING, JUDITH S.
 1991. *Amazon Crude.* New York: Natural Resources Defense Council.

KLUCKHOHN, CLYDE
 1951. "Values and Value-Orientations in the Theory of Action: An Exploration in Definition and Classification." In *Toward a General Theory of Action: Theoretical Foundations for the Social Sciences,* ed. T. Parsons and E. Shils, 388–433. First edition. Cambridge: Harvard University Press.

KOCKELMAN, PAUL
 2010. "Value Is Life Under an Interpretation: Existential Commitments, Instrumental Reasons and Disorienting Metaphors." *Anthropological Theory* 10(1–2):149–163.

KOHN, EDUARDO
 2002. "Infidels, Virgins, and the Black-Robed Priest: A Backwoods History of Ecuador's Montaña Region." *Ethnohistory* 49(3):545–582.

LATOUR, BRUNO

1993. *We Have Never Been Modern*. Cambridge: Harvard University Press.

1999. *Pandora's Hope: Essays on the Reality of Science Studies*. Cambridge: Harvard University Press.

LAUER, MATTHEW

2006. "State-led Democratic Politics and Emerging Forms of Indigenous Leadership among the Ye'kwana of the Upper Orinoco." *Journal of Latin American Anthropology* 11:51–108.

LEAR, JONATHAN

2006. *Radical Hope: Ethics in the Face of Cultural Devastation*. Cambridge: Harvard University Press.

LI, TANIA MURRAY

2000. "Articulating Indigenous Identity in Indonesia: Resource Politics and the Tribal Slot." *Comparative Studies in Society and History* 42(1):149–179.

2007. *The Will to Improve: Government, Development, and the Practice of Politics*. Durham: Duke University Press.

LITTLE, PAUL E.

1992. *Ecología política del cuyabeno: El desarrollo no sostenible de la Amazonía*. Quito: ILDIS, Abya-Yala.

LOHMANN, LARRY

1993. "Green Orientalism." *The Ecologist* 23(6):202–204.

LU HOLT, FLORA

2005. "The Catch-22 of Conservation: Indigenous Peoples, Biologists, and Cultural Change." *Human Ecology* 33(2):199–215.

LUKE, STEVEN

1999. "Environmentality as Green Governmentality." In *Discourses of the Environment*, ed. E. Darier, 121–151. Oxford: Blackwell.

MACDONALD, THEODORE, JR.

2002. "Ecuador's Indian Movement: Pawn in a Short Game or Agent in State Reconfiguration?" In *The Politics of Ethnicity: Indigenous Peoples in Latin American States*, ed. D. Maybury-Lewis, 168–198. Cambridge: The David Rockefeller Center for Latin American Studies, Harvard University.

MARCUS, GEORGE E., AND MICHAEL G. POWELL

2003. "From Conspiracy Theories in the Incipient New World Order of the 1990s to Regimes of Transparency Now." *Anthropological Quarterly* 76(2): 323–334.

MARTZ, JOHN D.

1987. *Politics and Petroleum in Ecuador*. New Brunswick, NJ: Transaction Publishers.

MARX, KARL

1977. *Capital*. Trans. B. Fowkes. Vol. 1. London: Penguin Books.

MEAD, GEORGE HERBERT

1934. *Mind, Self, and Society: From the Standpoint of a Social Behaviorist*. Chicago: University of Chicago Press.

MOORE, DONALD

2000. "The Crucible of Cultural Politics: Reworking 'Development' in Zimbabwe's Eastern Highlands." *American Ethnologist* 26(3):654–689.

MUEHLEBACH, ANDREA

2001. "'Making Place' at the United Nations: Indigenous Cultural Politics at the U.N. Working Group on Indigenous Populations." *Cultural Anthropology* 16(3):415–448.

2003. "What Self in Self-Determination? Notes from the Frontiers of Transnational Indigenous Activism." *Identities: Global Studies in Culture and Power* 10:241–268.

MUNN, NANCY

1986. *The Fame of Gawa: Value Transformation in a Massim (Papua New Guinea) Society.* Cambridge: Cambridge University Press.

NADASDY, PAUL

2005. *Hunters and Bureaucrats: Power, Knowledge, and Aboriginal-State Relations in the Southwest Yukon.* Seattle: University of Washington Press.

2006. "Transcending the Debate over the Ecologically Noble Indian: Indigenous Peoples and Environmentalism." *Ethnohistory* 52(2):291–331.

NIETZSCHE, FRIEDRICH

1992. "On the Genealogy of Morals." In *Basic Writings of Nietzsche*, ed. W. Kaufmann, 437–599. New York: Modern Library.

OINCE

1998. *El mejor lugar de la selva: Propuesta para la recuperación del territorio cofán.* Quito: Abya-Yala.

ORTIZ, SERGIO ELIAS

1954. *Notas sobre los indios kofán o kofane.* Bogotá: Ediciones Lerner.

OVERING, JOANNA

1989. "The Aesthetics of Production: The Sense of Community among the Piaroa and Cubeo." *Dialectical Anthropology* 14:159–175.

PARSONS, TALCOTT, AND EDWARD SHILS, EDS.

1994. *Toward a General Theory of Action: Foundations for the Social Sciences.* New Brunswick, NJ: Transaction Publishers.

PEEKE, M. CATHERINE

1973. *Preliminary Grammar of Auca.* Norman, OK: Summer Institute of Linguistics.

PEREIRA DE QUEIROZ, MARÍA

1969. *Historia y etnología de los movimientos mesiánicos.* Mexico City: Siglo Veintiuno Editores.

PINKLEY, HOMER V.

1973. "The Ethno-Ecology of the Kofan Indians." Ph.D. diss., Harvard University.

PORTER, THEODORE

1999. "Quantification and the Accounting Ideal in Science." In *The Science Studies Reader*, ed. M. Biagioli, 394–406. London: Routledge.

PUTNAM, HILARY

1975. *Mind, Language, and Reality.* Cambridge: Cambridge University Press.

RAMÍREZ, MARÍA C.

2002. "The Politics of Identity and Cultural Difference in the Colombian Amazon: Claiming Indigenous Rights in the Putumayo Region." In *The Politics of Ethnicity: Indigenous Peoples in Latin American States*, ed. D. Maybury-Lewis, 135–168. Cambridge: The David Rockefeller Center for Latin American Studies, Harvard University.

RAMOS, ALCIDA RITA

1994. "The Hyperreal Indian." *Critique of Anthropology* 14:153–171.

1998. *Indigenism: Ethnic Politics in Brazil.* Madison: University of Wisconsin Press.

RAPPAPORT, JOANNA

2005. *Intercultural Utopias: Public Intellectuals, Cultural Experimentation, and Ethnic Pluralism in Colombia.* Durham: Duke University Press.

REDFORD, KENT H.

1991. "The Ecologically Noble Savage." *Cultural Survival Quarterly* 15(1): 46–48.

REDFORD, KENT H., AND ALLYN M. STEARMAN

1993. "Forest-Dwelling Native Amazonians and the Conservation of Biodiversity: Interests in Common or Collision?" *Conservation Biology* 7(2):248–255.

RICHARDSON, KYLE

2003. "The Battle for Putumayo." *Cultural Survival Quarterly* (Winter):18–20.

RICOEUR, PAUL

1994. "Imagination in Discourse and Action." In *Rethinking Imagination: Creativity and Culture*, ed. G. Robinson and J. Rundell, 118–135. London: Routledge.

RIVAL, LAURA

1996. "Formal Schooling and the Production of Modern Citizens in the Ecuadorian Amazon." In *The Cultural Production of the Educated Person: Critical Ethnographies of Schooling and Local Practice*, ed. B. A. Levinson, D. E. Foley, and D. E. Holland, 153–167. Albany: State University of New York Press.

2002. *Trekking through History: The Huaorani of Amazonian Ecuador.* New York: Columbia University Press.

ROBINSON, SCOTT S.

1979. "Towards an Understanding of Kofan Shamanism." Ph.D. diss., Anthropology, Cornell University.

ROGERS, MARK

1996. "Beyond Authenticity: Conservation, Tourism, and the Politics of Representation in the Ecuadorian Amazon." *Identities: Global Studies in Culture and Power* 3(1–2):73–125.

RUTHERFORD, PAUL

1999. "The Entry of Life into History." In *Discourses of the Environment*, ed. E. Darier, 37–62. Oxford: Blackwell.

SAHLINS, MARSHALL

1985. *Islands of History.* Chicago: University of Chicago Press.

SALAZAR, ERNESTO

    1981. "The Federación Shuar and the Colonization Frontier." In *Cultural Transformations and Ethnicity in Modern Ecuador*, ed. N. E. Whitten, Jr., 589–613. Urbana: University of Illinois Press.

SAN SEBASTIÁN, MIGUEL, BEN ARMSTRONG, AND CAROLYN STEVENS

    2002. "Outcomes of Pregnancy among Women Living in the Proximity of Oil Fields in the Amazon Basin of Ecuador." *International Journal of Occupational Environmental Health* 8(4):312–319.

SAN SEBASTIÁN, MIGUEL, AND ANNA-KARIN HURTIG

    2004. "Oil Exploitation in the Amazon Basin of Ecuador: A Public Health Emergency." *Pan American Journal of Public Health* 15(3):205–211.

SANTOS-GRANERO, FERNANDO

    1992. "La sublevación mesiánica y anticolonial de Juan Santos Atahuallpa, 1742–1752." In *Etnohistoria de la alta Amazonía, Siglos XV–XVIII*, 237–258. Quito: Abya-Yala, CEDIME, MLAL.

SANTOS-GRANERO, FERNANDO, AND FREDERICA BARCLAY

    1998. *Selva Central: History, Economy, and Land Use in Peruvian Amazonia*. Washington, DC: Smithsonian Institution Press.

SAWYER, SUZANA

    2004. *Crude Chronicles: Indigenous Politics, Multinational Oil, and Neoliberalism in Ecuador*. Durham: Duke University Press.

SCHODT, DAVID W.

    1987. *Ecuador: An Andean Enigma*. Boulder: Westview Press.

SCHUTZ, ALFRED, AND THOMAS LUCKMANN

    1973. *The Structures of the Life-World*, Vol. 1. Trans. R. M. Zaner and J. H. Tristram Engelhardt. Evanston, IL: Northwestern University Press.

SERRANO, FERNANDO

    1993. "The Transformation of the Indian Peoples of the Ecuadorian Amazon into Political Actors and Its Effects on the State's Modernization Policies." Master's thesis, Anthropology, University of Florida.

SHAPIRO, JUDITH

    1987. "From Tupa to the Land without Evil: The Christianization of Tupi-Guarani Cosmology." *American Ethnologist* 14:126–139.

STERN, STEVE J.

    1993. *Peru's Indian Peoples and the Challenge of Spanish Conquest: Huamanga to 1640*. Madison: University of Wisconsin Press.

SWARTZ, SPENCER, AND MERCEDES ALVARO

    2010. "Ecuador Renegotiates with Foreign Oil Firms." *Wall Street Journal*, August 9.

SZEMINSKI, JAN

    1987. "Why Kill the Spaniard?: New Perspectives on Andean Insurrectionary Ideology in the 18th Century." In *Resistance, Rebellion, and Consciousness in the Andean Peasant World, 18th to 20th Centuries*, ed. S. J. Stern, 166–192. Madison: University of Wisconsin Press.

TAYLOR, CHARLES

    1985. *Human Agency and Language*. Cambridge: Cambridge University Press.

TIDWELL, MIKE

1996. *Amazon Stranger*. New York: Lyons & Buford.

TOWNSEND, WENDY, ET AL.

2005. "Cofán Indians' Monitoring of Freshwater Turtles in Zábalo, Ecuador." *Biodiversity and Conservation* 14(11):2743–2755.

TSING, ANNA LOWENHAUPT

2004. *Friction: An Ethnography of Global Connection*. Princeton: Princeton University Press.

TURNER, TERENCE

1984. "Value, Production and Exploitation in Simple Non-Capitalist Societies." Unpublished manuscript.

1988. "Commentary: Ethno-Ethnohistory: Myth and History in Native South American Representations of Contact with Western Society." In *Rethinking History and Myth: Indigenous South American Perspectives on the Past*, ed. J. D. Hill, 235–281. Urbana: University of Illinois Press.

1995. "An Indigenous Amazonian People's Struggle for Socially Equitable and Ecologically Sustainable Production: The Kayapó Revolt against Extractivism." *Journal of Latin American Anthropology* 1(1):98–121.

1999a. "Activism, Activity Theory, and the New Cultural Politics." In *Activity Theory and Social Practice*, ed. S. Chaikin, M. Hedegaard, and U. J. Jensen, 114–135. Aarhus, Denmark: Aarhus University Press.

1999b. "Indigenous and Culturalist Movements in the Contemporary Global Conjuncture." In *Las identidades y las tensiones culturales de modernidad*, 53–72. Santiago de Compostela, Spain: Federación de Asociaciones de Antropología del Estado Español.

2000. "Indigenous Rights, Environmental Protection and the Struggle over Forest Resources in the Amazon: The Case of the Brazilian Kayapó." In *Earth, Air, Fire and Water: The Humanities and the Environment*, ed. J. Conway, Kenneth Thompson, and Leo Marx, 145–169. Boston: University of Massachusetts Press.

2002. "Shifting the Frame from Nation-State to Global Market: Class and Social Consciousness in the Advanced Capitalist Countries." *Social Analysis* 46(2):56–80.

2003a. "The Beautiful and the Common: Inequalities of Value and Revolving Hierarchy among the Kayapó." *Tipití: Journal of the Society for the Anthropology of Lowland South America* 1(1):11–26.

2003b. "Class Projects, Social Consciousness, and the Contradictions of 'Globalization.'" In *Violence, the State and Globalization*, ed. J. Friedman, 35–66. New York: Altamira.

2008. "Marxian Value Theory: An Anthropological Perspective." *Anthropological Theory* 8(1):43–56.

UNGER, ROBERTO MANGABEIRA

1987. *Social Theory: Its Situation and Its Task*. Cambridge: Cambridge University Press.

UQUILLAS, JORGE E.

1993. "La tenencia de la tierra en la Amazonia ecuatoriana." In *Retos de la Amazonia*, Teodoro Bustamante et al., 61–94. Quito: ILDIS, Abya-Yala.

USAID

2005a. *Ecuador: Biodiversity Conservation.* 518–001. Washington, DC.

2005b. *Budget: Ecuador.* Washington, DC.

UZENDOSKI, MICHAEL

2005. *The Napo Runa of Amazonian Ecuador.* Urbana: University of Illinois Press.

VARESE, STEFANO

1996. "The New Environmentalist Movement of Latin American Indigenous People." In *Valuing Local Knowledge: Indigenous People and Intellectual Property Rights*, ed. Stephen B. Brush and Doreen Stabinsky, 122–142. Washington, DC: Island Press.

VARGAS, RICARDO

2004. "State, Esprit Mafioso, and Armed Conflict in Colombia." In *Politics in the Andes: Identity, Conflict, Reform*, ed. J.-M. Burt and P. Mauceri, 107–125. Pittsburgh: University of Pittsburgh Press.

VEBER, HANNE

1998. "The Salt of the Montaña: Interpreting Indigenous Activism in the Rain Forest." *Cultural Anthropology* 13(3):382–413.

2003. "Asháninka Messianism: The Production of a 'Black Hole' in Western Amazonian Ethnography." *Current Anthropology* 44(2):382–413.

2007. "Merits and Motivations of an Ashéninka Leader." *Tipití: Journal of the Society for the Anthropology of Lowland South America* 5(1):9–23.

VIATORI, MAXIMILIAN

2007. "Zápara Leaders and Identity Construction in Ecuador: The Complexities of Indigenous Self-Representation." *Journal of Latin American and Caribbean Anthropology* 12(1):104–133.

VICKERS, WILLIAM T.

2003. "The Modern Political Transformation of the Secoya." In *Millennial Ecuador: Critical Essays on Cultural Transformations and Social Dynamics*, ed. N. E. Whitten, Jr., 46–74. Iowa City: University of Iowa Press.

VIVEIROS DE CASTRO, EDUARDO

1992. *From the Enemy's Point of View: Humanity and Divinity in an Amazonian Society.* Trans. C. Howard. Chicago: University of Chicago Press.

WALCOTT, JUDITH

2003. "Nowhere Left to Run: An Indigenous Ecuadorian Perspective on Plan Colombia." *Cultural Survival Quarterly* (Winter):58–59.

WARREN, KAY B., AND JEAN E. JACKSON

2002. *Indigenous Movements, Self-Representation, and the State in Latin America.* Austin: University of Texas Press.

2002. "Introduction: Studying Indigenous Activism in Latin America." In *Indigenous Movements, Self-Representation, and the State in Latin America*, ed. K. B. Warren and J. E. Jackson, 1–46. Austin: University of Texas Press.

WAYNE, E. ANTHONY

2004. "State's Wayne Testifies on U.S. Investment in Peru, Ecuador." Distributed by the Bureau of International Information Programs, U.S. Department of State.

WEBBER, JEFFREY R.

2010. "Ecological Resistance, Indigenous Struggle, and Rafael Correa's Neo-Extractivism in Ecuador: An Interview with Gloria Chicaiza." *New Socialist: Ideas for Radical Change*, July 14. http://www.newsocialist.org/index .php?option=com_content&view=article&id=234:ecological-resistance -indigenous-struggle-and-rafael-correas-neo-exctractivism-in-ecuador-an -interview-with-gloria-chicaiza-&catid=51:analysis&Itemid=98.

WEST, PAIGE

2006. *Conservation Is Our Government Now: The Politics of Ecology in Papua New Guinea*. Durham: Duke University Press.

WHITTEN, NORMAN E., JR.

1976. *Sacha Runa: Ethnicity and Adaptation of Ecuadorian Jungle Quichua*. Urbana: University of Illinois Press.

1981. "Amazonia Today at the Base of the Andes: An Ethnic Interface in Ecological, Social, and Ideological Perspectives." In *Cultural Transformations and Ethnicity in Modern Ecuador*, ed. N. E. Whitten, Jr., 1–41. Urbana: University of Illinois Press.

1985. *Sicuanga Runa: The Other Side of Development in Amazonian Ecuador*. Urbana: University of Illinois Press.

2003. Introduction. In *Millennial Ecuador: Critical Essays on Cultural Transformations and Social Dynamics*, ed. N. E. Whitten, Jr., 1–45. Iowa City: University of Iowa Press.

2004. "Ecuador in the New Millennium: 25 Years of Democracy." *The Journal of Latin American Anthropology* 9(2):439–460.

WHITTEN, NORMAN E., JR., AND DOROTHEA SCOTT WHITTEN

2008. *Puyo Runa: Imagery and Power in Modern Amazonia*. Urbana: University of Illinois Press.

WRIGHT, ROBIN

1988. "Anthropological Presuppositions of Indigenous Advocacy." *Annual Review of Anthropology* 17:365–390.

2002. "Prophetic Traditions among the Baniwa and other Arawakan Peoples in the Northwest Amazon." In *Comparative Arawakan Histories*, ed. J. Hill and F. Santos-Granero, 269–293. Urbana: University of Illinois Press.

WRIGHT, ROBIN, AND JONATHAN HILL

1986. "History, Ritual, and Myth: Nineteenth-Century Millenarian Movements in Northwest Amazonia." *Ethnohistory* 33:31–54.

1992. "Venancio Kamiko: Wakuénai Shaman and Messiah." In *Portals of Power: Shamanism in South America*, ed. E. Jean Matteson Langdon and Gerhard Baer, 257–286. Albuquerque: University of New Mexico Press.

YOUNGERS, COLETTA A.

2004. "Collateral Damage: The U.S. 'War on Drugs' and Its Impact on Democracy in the Andes." In *Politics in the Andes: Identity, Conflict, Reform,* ed. J.-M. Burt and P. Mauceri, 126–145. Pittsburgh: University of Pittsburgh Press.

ŽIŽEK, SLAVOJ

2009. *In Defense of Lost Causes.* London: Verso.

Index

Page numbers in italics indicate photos.

Chiritza, 31, 37, 39

Christian and Missionary Alliance, 182

Clinton, Bill, 11; and Plan Colombia, 109

coadministration, 112, 113, 114, 115

Coca (town), 71, 228ch6n11

coca crops, 11, 12, 21, 96, 126–127; fumigation of, 11

Coca River, 186

cocaine, 10, 126–127. *See also* drug trade

Cocama, 58, 63, 64, 92, 95, 112, 120, 128, 134, 144, 193, 219, 224ch2n4; as Cofán leaders, 71; and Cofán-ness, 67; and Cofán students, 180, 181; and Cofán women, 194, 195; Colombian, 129; intermarriage of, with Cofán, 59, 195, 196; view of Cofán, 124, 125, 184. *See also* mestizos

cocoya (malevolent supernatural agents), 13, 89, 90, 94, 149, 210, 213; appearance of, 92; use of term in Colombia, 13, 93. *See also* shamanism; *yaje a'i* (supernatural people)

Cofán-Bermejo Ecological Reserve (RECB), 110, 111, 112, 113, 115, 166; management of, 114

Cofán case, 80, 93, 98, 114, 217, 218, 219, 221; and goal of conservation, 222

Cofanes River/Río Cofanes, 7, 14, 57

Cofán ethnic federation, 14, 21, 22, 24, 43, 94, 104, 197. *See also* Federación Indígena de la Nacionalidad Cofán del Ecuador (FEINCE); FSC

Cofán experiment, 2, 25, 54, 103, 185, 199, 207, 214, 215, 219; as utopian moment, 223. *See also* Cofán case

Cofán Historical Mapping Project, 24, 205, 206

Cofán-ness, 5, 12, 56, 61–69, 77, 94, 181, 197, 208; Aguinda's, 121; and *ai've dambi'choa*, 68; Borman's,

43, 56, 73–77; complicating of, 71–72; and following Cofán norms, 62; and intermarriage, 196; levels of, 67; and participation in the struggle, 98–99; process leading to, 69–70; questioning of, 83. *See also a'i*; *opa*-ness

Cofán park guards, 96, 103, 105, 111, 112, 113, 201, 203, 214, 225ch3n9. *See also* Cofán Ranger Program

Cofán people, x, 1, 7, 15, 16, 18, 50, 88, 89, 111, 119, 139, 141, 189, 216, 219–220, 223; and acceptance of Borman, 41, 44, 56, 66, 67, 73, 78, 209, 212; and anxiety, 2, 19, 51, 81, 84, 85, 86, 87, 143, 216; and appreciation of forest, 146–147; athletic abilities of, 180, 228ch6n5; avoiding ethnocide, 15; challenges of, 208; and Cocama, 124, 125, 128, 129, 144; in Colombia, 5, 10, 112; conservationism of, 219–222; contemporary situation of, 8–15; and cooperation, 142, 143, 164; and cultural change, 195–196; customs of (*a'i canse' cho*), 63, 67, 69, 72, 73, 74, 76, 83, 84, 88, 121, 136, 140, 141–142, 149, 171–172, 174, 190, 227ch5n5; dress of, 60, 63; and ethnic ambiguity, 25, 70–73, 78, 133, 175, 177, 207, 213; and ethnic exclusion, 5, 12, 62, 63, 65; ethnicity of, 58–61; and ethnic others, 5, 58–61, 65, 71, 82, 91, 92, 94, 96, 104, 121, 164, 178, 194, 195, 218, 224ch2n2; ethnonyms of, 5–6; and fear, 149, 162, 174, 187; fearlessness of, 87, 93, 94, 96, 97; food of (*cui'ccu*), 63, 79, 83, 188, 189, 190, 226ch5n4; and forest-based lifestyle, 64, 65, 66, 73, 107, 111, 142, 147, 208; and FSC, 116–120, 121, 127, 128, 133–134, 135; and genealogy, 61–62; and gossip, 117, 124, 135, 152, 162, 202; historians of, 6; history of,

Cuyabeno River, 7
Cuyabeno Wildlife Reserve, 22, 114,
147, 148, 155

davu, 89, 90, 91, 92–93. See also
curaga (shaman); shamanism
deforestation, 9, 10. *See also* environmental destruction
Detore, Miranda, 203–204, 205
Dewey, John, 17
Díaz de Pineda, Gonzalo, 7
disease, 6, 7, 8, 15, 213
Doreno (Dureno) community, 8, 13, 15,
30, 31, 38, 71, 91, 110, 140, 144,
148, 188; A'ingae-language school
in, 175–176; Borman's return to, 34,
36; and care of the forest (*tsampima
coiraye*), 144; and colonization, 36;
comuna status of, 38; and oil industry, 21, 36; and social change, 196;
and Texaco, 9
Dovuno community, 15, 41, 68, 111,
192; and Napo intermarriage, 195
drug trade, 10, 15; production of, 11,
13; war on, 11. *See also* coca crops;
cocaine
Dureno Uno (oil well), 9
du'shu (child). *See* childhood
dusunga (unmarried sexually mature
person), 171, 173, 194; female, 172;
male, 173

ecchoen'cho (mixed [ethnicity]), 58–59,
62. *See also* Cocama; Cofán people,
and ethnic others; mestizos
Ecocanoas/Ecocanoes, 105, 121
ecological reserves. *See* protected areas;
territorial protection
ecotourism, 2, 38, 39, 109, 138; in
Sucumbíos, 12; in Zábalo, 145. *See
also* tourism
Ecuador, 13; army of, 12; corruption
in, 109, 125; functional illiteracy
in, 128; and indigenous, 13–14;
national education system of, 37;
1998 constitution of, 14; and taxes,

125, 128, 129, 130; U.S. investors
in, 125; violence in, 105
Ecuadorian Institute of Agrarian
Reform and Colonization, 22, 36
English Fellowship Church, 45, 186
environmental conservation 1, 2; and
FSC, 107; and local peoples, 219–
222. *See also* conservation
environmental destruction, ix, xii, 1, 4,
9, 10, 38, 53, 138, 203, 214
environmental knowledge, 142, 143,
159, 160. *See also* forest knowledge
environmental politics, 103, 107,
110–115, 208. *See also* Cofán
people, conservationism of; conservation; Field Museum of Natural
History, and Environmental and
Conservation Programs; leadership;
NGOs
Escuela del Futuro, 178
Espinosa, Freddy, 106, 107, 117, 119,
121–122, 123, 129, 179, 191; salary
of, 226ch4n3; view of Cofán, 124
ethnic difference, 61, 96, 116, 120, 181
ethnic federations, 13–14, 207. *See also*
Cofán ethnic federation
ethnic identity, 58–61
ethnography, 1, 18, 19, 20, 21; holistic,
19, 20; possibilities of, 222–223
European Community, 106
extractivism, 15. *See also* oil industry

facturas, 127–129, 131, 132. *See also*
transparency
Federación Indígena de la Nacionalidad
Cofán del Ecuador (Indigenous
Federation of the Cofán Nationality
of Ecuador; FEINCE), 14, 44, 71,
104, 110, 115, 120, 166, 197, 206;
and Acta de Acuerdos, 114; and
Aguinda, 121; and FSC, 105, 110,
113
Federation of Shuar Centros, 13. *See
also* indigenous movement
Ferrer, Rafael (Jesuit missionary), 6,
7, 95

Field Museum of Natural History, 23, 105, 121, 124, 201, 203; accounting practices of, 130; and Center for Cultural Understanding and Change (CCUC), 227ch5n10; and Cofán Historical Mapping Project, 205; and Division of Environment, Culture, and Conservation (ECCo), 227ch5n10; and Environmental and Conservation Programs (ECP), 155–166, 167, 205, 227ch5nn10–11; and funding FSC, 126; and Zábalo, 133, 134, 138, 221

Flotel Orellana, 145, 178. *See also* tourism; Zábalo community

forest knowledge, 37, 42, 107, 145–146, 167, 190, 191

Forestry Law (Ecuador), 113

Foucault, Michel, and "governmentality," 221–222

Foundation for the Survival of the Cofán People. *See* FSC

Friction: An Ethnography of Global Connection (Tsing), ix, x

Friede, Juan, 6, 7

FSC, 14, 24, 71, 76, 103, 125, 166, 168, 192, 203, 214, 220, 225ch4n1, 226ch4n4; accounting practices of, 127–132, 226ch4n6; and *a'i*, 122, 123, 127; and Alliance Academy, 228ch6n7; Cofán understanding of, 134; and Cofán vision, 115; and Colegio Francés, 180, 184; creation of, 104, 105; and Education Project, 170, 178, 185, 192, 195, 203, 206–207, 209, 215, 228ch6nn4,10; and FEINCE, 105, 110, 113; funding of, 105–106, 109, 110, 118–119, 125, 126, 130; growth of, 116; and IOM, 131; lack of confidence at, 125; mission of, 107; and mobilization, 121, 137; and salaries, 226ch4n3; staff of, 106–107, 117, 122, 127, 130, 132, 133, 134, 136, 180; workplace of, 116–117

fuite'cho (help), 134

Fundación Interayuda, 178. *See also* Tumbaco school

Fundación Natura (NGO), 44

gender, 90, 141, 143, 172, 173, 192–193, 194

Global Conservation Fund, 106

globalization, xi, 4, 21, 206

global vision, 15, 18, 19, 26, 78, 107, 109, 110, 115, 166–168, 181, 219, 221

gold mining, 8

governmentality, 16, 221–222

Government of Peoples and Nationalities, 15. *See also* Confederation of Indigenous Nationalities of Ecuador

Graeber, David, xi

Gramsci, Antonio, x

grants, 118, 119, 122, 130; and Borman, 48, 66, 96, 106, 120; and FSC, 104, 105, 107, 118; and funding crises, 119, 226ch4n3; and ICCA, 226ch4n2

"gringo chief," 4, 54, 170; eventual replacement of, 206. *See also* Borman, Randy

gringos, 15, 44, 64, 92, 134, 219; and appreciation of Cofán, 183–184; arrival of, 59; Borman as, 75, 76, 121; term of, as insult, 65

Guamués River, 7, 212; valley of, 10

Guatemala, 31, 37

Güeppí River, 22

guerrillas, 11, 12, 21, 93. *See also* National Liberation Army; Revolutionary Armed Forces of Colombia (FARC)

hallucinogens, 19, 60, 88, 174; *yaje*, 88, 89, 90, 97, 140, 174, 193. *See also* shamanism

Harlow, Chip, 105, 117

Harvey, David, ix; as "insurgent architect," x

Heidegger, Martin, 17, 190

yaje a'i (supernatural people), 88, 89, 90, 149, 174. *See also* shamanism

Yiyoguaje, Nivaldo, 205; educational path of, 201–203

Zábalo community, 1, 3, 12, 23, 110, 142, 145, 154, 167, 183; compared to Quito, 187–192; as environmental experiment, 138; and Field Museum, 133, 134, 138, 221; formation of, 3, 9, 22–23, 31, 37–38, 39, 144, 215, 224ch1n5, 226ch5n3; and poverty, 140; publicity of, 3; and resistance to oil industry, 43, 55, 156, 224ch1n3; schooling in, 175–178, 201; state recognition of, 15, 22; and tourism, 105, 109, 117, 123, 145

Zábalo residents, 13, 16, 21, 22, 68, 82, 116, 120, 134, 139, 142, 143, 154, 165, 167; Antonio and Pablo, 79–80, 85, 87, 92–93, 225ch3n1; and *a'qquia injanga* ways, 153; on Borman, 73–77, 170; and care of forest, 146, 147, 153; and Cofán identity, 18, 65, 199, 200; and communal prohibitions (*se'pi'cho*), 148–155, 160, 163, 226ch5n4, 227ch5nn5,6,8, 228ch5n14; and conservation, 19, 103, 138, 144; and Cuyabeno Wildlife Reserve, 23, 147, 148; and ECP projects, 156–158, 159, 161, 162, 164, 165, 166, 227ch5n11, 227ch5n12, 228ch5n15; and FSC, 133–134, 136; on Napo villages, 82; on outsider help (*fuite'cho*), 134–135, 165; on outside schooling, 197–198; and protection of territory, 111, 112; and subsistence practices, 146, 147, 153; and tourism, 146, 148

Žižek, Slavoj, xi

CPSIA information can be obtained at www.ICGtesting.com
Printed in the USA
LVOW12s0902280115

424591LV00003B/209/P